Building iPhone and iPad
Electronic Projects

Mike Westerfield

O'REILLY®

Beijing · Cambridge · Farnham · Köln · Sebastopol · Tokyo

Building iPhone and iPad Electronic Projects

by Mike Westerfield

Copyright © 2013 James M. Westerfield. All rights reserved.

Printed in the United States of America.

Published by O'Reilly Media, Inc., 1005 Gravenstein Highway North, Sebastopol, CA 95472.

O'Reilly books may be purchased for educational, business, or sales promotional use. Online editions are also available for most titles (*http://my.safaribooksonline.com*). For more information, contact our corporate/institutional sales department: 800-998-9938 or *corporate@oreilly.com*.

Editor: Courtney Nash	**Indexer:** WordCo Indexing Services
Production Editor: Melanie Yarbrough	**Cover Designer:** Randy Comer
Copyeditor: Rachel Head	**Interior Designer:** David Futato
Proofreader: Linley Dolby	**Illustrator:** Rebecca Demarest

September 2013:   First Edition

Revision History for the First Edition:

2013-09-10:   First release

See *http://oreilly.com/catalog/errata.csp?isbn=9781449363505* for release details.

ISBN: 978-1-449-36350-5

LSI

Table of Contents

Preface

You carry an amazing scientific instrument around in your pocket every day, using it for mundane tasks like making phone calls or listening to music. Your iPad 2 is as fast as a Cray-2 supercomputer from just a few decades ago, yet most people only use it to read books or surf the Web. What a waste.

This book is all about connecting your iPhone, iPod Touch, or iPad to the real world. You'll start by learning how to access the sensors built right into your device. Next you'll see how to connect wired sensors through the headphone port using a wonderful little device called HiJack. Several chapters show various ways to use Bluetooth low energy to connect to sensors, Arduino microcontrollers, motor controllers, and even other iPhones or iPads. Finally, you'll see exactly how to use WiFi to connect to the Internet or physical devices connected to WiFi devices.

It would be pretty boring to make all of these connections just to make a few LEDs light up, so the book is organized around fun, interesting projects. The built-in sensors are used to create a metal detector. HiJack is hooked up to a simple electrical device so it can be used as a plant moisture sensor. Bluetooth low energy connects to a Texas Instruments SensorTag to detect acceleration to track the flight of a model rocket, and later to an Arduino microcontroller to hack a radio-controlled car, showing how to create robots and control them with your iPhone. Bluetooth low energy can also be used for peer-to-peer communication between iOS devices. You will learn how this is done by creating an arcade game that uses iPhones for game paddles. WiFi will be hooked up to a serial bridge to control servos, ultimately hacking a candy dispenser to give you candy under iPhone control.

Our look at each topic starts with a chapter that introduces the basic concepts using a simple project. One or more chapters follow these introductions, presenting the fun projects just mentioned. You may not want to build every one of them yourself, but reading through how they are created and how they work, you will get ideas about how to build your own projects.

You don't need to go through this book linearly. If a project in the middle of the book seems really interesting, jump right to it. Each chapter starts with a section called "About This Chapter." It lists the prerequisites, telling you which other chapters contain information you might need before attempting the project in the chapter you are interested in.

All of the hardware in the book is developed with electronic components you can buy from many Internet stores, but some of it is hard to find locally. Plan ahead. Glance at the parts list in a chapter a week or two before you want to get started, and order the parts you need.

Finally, the projects in this book cover several disciplines. There's a lot of software, quite a bit of electronics, and a fair amount of mechanical engineering involved. Some of the stuff in this book is going to seem beyond your abilities. I know a few of the projects seemed that way to me as I wrote the book. After all, even though most of us have some technical ability, either through education or experience with hobbies, almost no one is fully qualified at computer science, electrical engineering, mechanical engineering, and physics.

Be brave, grasshopper.

Everything is laid out very carefully. If you don't know much about software, start with the completely developed programs in the book, all of which are built right into techBASIC. If you don't know one end of a battery from another, just wire stuff as you see it in the diagrams and photos that carefully document each circuit. As you learn more, you can experiment. Sure, there will be some failures along the way. I burned out a circuit or two and crashed a lot of software writing the book, and you'll do the same as you read it. That's how we learn.

I hope you don't just build the projects in this book, though. The whole point is to learn *how* to do things, not just follow some plans. Whether you're a professional trying to figure out how to remotely access data from a buried seismograph, a student exploring robotics for a science fair project, or an inventor tinkering with awesome ideas in your garage, I hope this book gives you some techniques and ideas that will enable you to create amazing things by combining software, electronics, and mechanics to build devices.

So, let's go forth and control our world!

Conventions Used in This Book

The following typographical conventions are used in this book:

Italic
> Indicates new terms, URLs, email addresses, filenames, and file extensions.

`Constant width`

> Used for program listings, as well as within paragraphs to refer to program elements such as variable or function names, databases, data types, environment variables, statements, and keywords.

`Constant width bold`

> Shows commands or other text that should be typed literally by the user.

`Constant width italic`

> Shows text that should be replaced with user-supplied values or by values determined by context.

 This icon signifies a tip, suggestion, or general note.

 This icon indicates a warning or caution.

Using Code Examples

This book is here to help you get your job done. Where this book includes code examples, you may use the code in this book in your programs and documentation. You do not need to contact us for permission unless you're reproducing a significant portion of the code. For example, writing a program that uses several chunks of code from this book does not require permission. Selling or distributing a CD-ROM of examples from O'Reilly books does require permission. Answering a question by citing this book and quoting example code does not require permission. Incorporating a significant amount of example code from this book into your product's documentation does require permission.

We appreciate, but do not require, attribution. An attribution usually includes the title, author, publisher, and ISBN. For example: "*Building iPhone and iPad Electronic Projects* by Mike Westerfield (O'Reilly). Copyright 2013 James M. Westerfield, 978-1-449-36350-5."

If you feel your use of code examples falls outside fair use or the permission given above, feel free to contact us at *permissions@oreilly.com*.

Safari® Books Online

Safari Books Online is an on-demand digital library that delivers expert content in both book and video form from the world's leading authors in technology and business.

Technology professionals, software developers, web designers, and business and creative professionals use Safari Books Online as their primary resource for research, problem solving, learning, and certification training.

Safari Books Online offers a range of product mixes and pricing programs for organizations, government agencies, and individuals. Subscribers have access to thousands of books, training videos, and prepublication manuscripts in one fully searchable database from publishers like O'Reilly Media, Prentice Hall Professional, Addison-Wesley Professional, Microsoft Press, Sams, Que, Peachpit Press, Focal Press, Cisco Press, John Wiley & Sons, Syngress, Morgan Kaufmann, IBM Redbooks, Packt, Adobe Press, FT Press, Apress, Manning, New Riders, McGraw-Hill, Jones & Bartlett, Course Technology, and dozens more. For more information about Safari Books Online, please visit us online.

How to Contact Us

Please address comments and questions concerning this book to the publisher:

O'Reilly Media, Inc.
1005 Gravenstein Highway North
Sebastopol, CA 95472
800-998-9938 (in the United States or Canada)
707-829-0515 (international or local)
707-829-0104 (fax)

We have a web page for this book, where we list errata, examples, and any additional information. You can access this page at *http://www.oreil.ly/building-iphone-ipad*.

To comment or ask technical questions about this book, send email to *bookquestions@oreilly.com*.

For more information about our books, courses, conferences, and news, see our website at *http://www.oreilly.com*.

Find us on Facebook: *http://facebook.com/oreilly*

Follow us on Twitter: *http://twitter.com/oreillymedia*

Watch us on YouTube: *http://www.youtube.com/oreillymedia*

Acknowledgments

When I was a young nerd toting my slide rule back and forth to the library, one of my favorite books was *The Amateur Scientist*, a collection of articles from *Scientific American*. It was a remarkably diverse collection of projects. I added a significant amount of wear to that book, and eventually bought and wore out my own copy.

I hope this book is a lot like that one—it's a book of projects, some of which you're unlikely to take the time to build yourself. I hope you wear it out thumbing through the pages. As you do, though, keep in mind that it's not the work of a single person. Oh, sure, I wrote it, but as Newton famously remarked, "If I have seen further it is by standing on the shoulders of giants."

I owe a great deal to the people who educated me, both in and out of the classroom. A lot of them were in the early Apple II community. I won't even try to name them, but you can find their footprints all through this book. Check out the KansasFest archives to meet some of these astoundingly creative people.

My wife is an amazing person. She's my cheerleader, my critic, and the first person to read and correct each page. She watched our house as it was taken over by rockets, robot cars, and remote-controlled gadgets, encouraging me without complaining about the mess. She even pitched in on many of the projects. Among other things, the eyeball in Chapter 11 is her artwork. What an amazing best friend.

Thomas Schmid from the University of Utah took the time to answer a lot of questions about the HiJack, no doubt keeping me from frying a few. Like a lot of components, HiJack is manufactured by Seeed Studio. Leslie Liao from Seeed Studio kindly supplied the book's reviewers with HiJacks so they could try the projects in Chapter 4 and Chapter 5.

I have some great new Internet friends at the Texas Instruments facility in Norway. Jarle Bøe was fantastic, getting me started with the SensorTag before it even came out. He also let me use some of his photos, which are credited in the text. His staff was more than just helpful—Jomar Hoensi even wrote a special version of the firmware so it could collect data up to ±8G for rocket flights, and took the time to answer a lot of neophyte questions as I came up to speed on Bluetooth low energy. The rockets you see in Chapter 7 exist because of their efforts. I'm happy to say the rockets got to go to Norway for some trade shows, even if I never made it there myself.

My reviewers patiently slogged through all or part of this book. The amazing and talented Ryan family made up most of the reviewers. Kevin Ryan, Jess Finley, and Ken Moreland spent countless hours making sure everything worked and the descriptions were clear enough to follow. They even had electronics parties where they got together to build the projects. Doyle Maleche joined, from afar, bringing his experience as an educator to bear on the book. I even got to get acquainted with a great O'Reilly author,

Alasdair Allan, who took the time to review parts of the book. Their comments made this a much better book than it would otherwise have been.

I've done a lot of writing for magazines over the years, and published software with a number of companies. While this is my first traditional book, I've worked with publishers and editors for a long time. I was pretty lucky to get some early training and encouragement from the editors and writers at Call A.P.P.L.E. I had pretty much given up on finding a publisher that really cared that much about its authors and products, but O'Reilly sure seems to be another one. I've been fortunate to have two great editors on this book. Brian Jepson got me started, then handed me off to Courtney Nash when Make: split from O'Reilly. Finding two people of their quality in a row says a lot for this company. If you decide to write, be sure to drop them a line. They are good people.

So, to all of you, from the Apple II buds in my early years to my newest friends at O'Reilly, thanks for making me look good!

Credits

While modified for the book, Chapter 5 originally appeared in the June 2012 issue of Nuts & Volts Magazine (*http://www.nutsvolts.com*) and is reprinted by permission of T & L Publications, Inc.

The SensorTag photo from Chapter 6 is courtesy of Jarle Bøe at Texas Instruments.

The illustration of the declination of the Earth's magnetic field in Chapter 3 is courtesy of Wikimedia Commons.

CHAPTER 1

Getting Familiar with techBASIC and Built-in Sensors

About This Chapter

Prerequisites
You should already be familiar with using your iPhone. You should have some idea what programming is all about, although you do not need to be an ace programmer. It helps to know some variant of BASIC, but this is not required.

Equipment
You will need an iPhone, iPod Touch, or iPad running iOS 5 or later.

Software
You will need a copy of techBASIC or techBASIC Sampler.

What You Will Learn
This chapter starts with an introduction to techBASIC, the technical programming language used in this book for accessing sensors. It shows how to access the accelerometer that is built into every iOS device, starting with a simple one-line program and working up to a sophisticated accelerometer app.

Your Own Tricorder

I was always a little jealous when Spock pulled out his tricorder on *Star Trek* and began measuring practically every physical value you could imagine. It's staggering how far technology has come, though. I carry a tricorder around in my pocket all the time now! Mine measures acceleration, rotation, and magnetic fields, giving both the strength and direction of each. It's not quite as sophisticated as Spock's, but it's also not so large and clunky.

This book is all about using your iPhone and iPad to control electronic devices, often sensors. We're going to start off with the sensors that are built right in, so you can pop out your tricorder and measure stuff, too.

The iPod Touch

The iPod Touch is essentially an iPhone without the phone, or, depending on your viewpoint, an iPad with a pocket-sized screen. Nothing in this book uses the phone part of the iPhone, so any time you see the iPhone mentioned, you can also use an iPod Touch. For the most part, we won't talk about the iPod Touch specifically, but keep in mind that you can always use one instead of an iPhone.

A Crash Course in techBASIC

We'll get started on the first instrument for our tricorder in a moment. First, though, let's take a look at the language we'll use for programming.

Our programs will be developed in a technical programming language called techBASIC, available in the App Store. There are a number of reasons for using techBASIC instead of Objective C, the programming language used to write most apps (including techBASIC). Here are some of the big ones:

- techBASIC runs right on your iPhone or iPad. You don't have to use, or even own, a Macintosh computer to write or run these programs.
- techBASIC is less expensive. While Xcode (the development environment for Objective C) is free, you must join Apple's developer program to actually move programs to your iOS device. That costs $99 every year. techBASIC costs $14.99 one time.
- techBASIC is simpler. It's designed specifically for writing technical programs and connecting to external devices. Programs that would take a half-dozen to a dozen classes, each with a half-dozen to a dozen methods, can often be written with just a few lines in techBASIC.
- techBASIC is easier to learn and more forgiving than Objective C, so you can concentrate on the fun part—writing the programs to control the Internet of Things.

techBASIC Sampler

There is a free version of techBASIC called techBASIC Sampler. It's also displayed in some places with the shortened name techSampler. The free version lets you view and run all of the samples that come with techBASIC. All of the programs from this book

---

are samples in techBASIC, so you can use the free version to run the programs. You can even use the debugger to trace through the programs. The only limitation is editing. techBASIC Sampler doesn't let you change a program or create a new one. When you try to edit a program or create a new one, techBASIC Sampler will invite you to upgrade to the full version of techBASIC using an in-app purchase. You can do that or just buy techBASIC.

That said, there are a couple of places in the book where we will create very short programs that are not preloaded as samples. These are typically used to show a simple feature before it gets buried in a longer program or to show how to use techBASIC. You can skip entering those programs without missing anything significant.

We'll just talk about techBASIC in the book, but other than editing, you can always perform the same tasks with techBASIC Sampler.

Where to Get techBASIC

Like all iOS apps, techBASIC and techBASIC Sampler are available from Apple's App Store.

- techBASIC (*http://itunes.apple.com/us/app/techbasic/id470781862?ls=1&mt=8*) is the full version of the development environment. It includes the samples from this book.
- techBASIC Sampler (*https://itunes.apple.com/us/app/techsampler/id626214040?ls=1&mt=8*) (also called techSampler, so the name will show up under the icon on the iPhone and iPad) is the free version of techBASIC. It allows you to run programs, including the samples from this book, but you cannot edit existing programs or create new ones. There is an in-app purchase to enable editing, which makes this program feature-for-feature compatible with techBASIC.

You can find more information about both programs, as well as a technical reference manual, at the Byte Works website (*http://www.byteworks.us/Byte_Works/Products.html*).

Running Your First Program

Crank up techBASIC and you will see a display something like Figure 1-1, depending on the device you are using. If you are using an iPad and holding it in portrait view, tap the Programs button at the top left of the screen to see the list of programs. The iPhone will start off showing the programs, but if you switch to another display, you can switch back by tapping the Programs button at the lower left of the iPhone display.

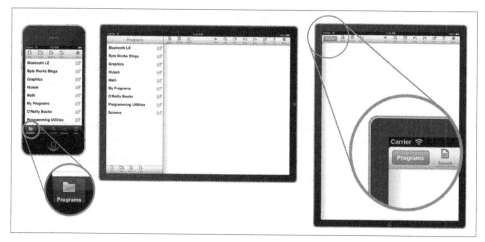

Figure 1-1. Initial techBASIC display

The program list shows all of the sample programs, contained in folders used to organize them. One of the folders is called *O'Reilly Books*, as shown in Figure 1-2. Tap the name and it will expand to show the programs in the folder. Tap the Programs navigation button at the top of the screen and the folder closes, moving you back to the original list of folders.

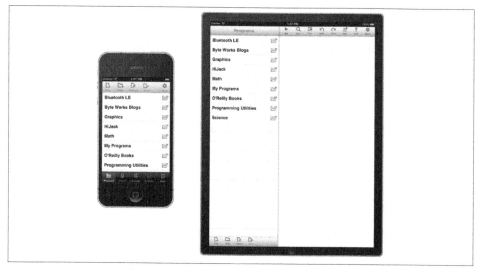

Figure 1-2. The O'Reilly Books folder

One of the programs is called Accelerometer. Tap the Edit button next to the program's name to see the source code for the program. On the iPhone, you can tap the Programs button to get back to the program list.

You run a program by tapping on the name of the program, as shown later in Figure 1-7. Give the accelerometer a try. The display will shift to the graphics screen, where you'll see a constantly updating plot of the acceleration experienced by the iPhone or iPad, as shown in FIG 1-4. The accelerometer is the first sensor for our tricorder, and we'll dig into the program in detail in a moment. Stop the program by tapping the Quit button.

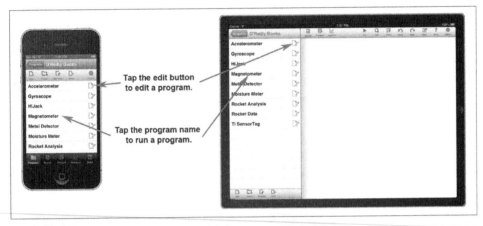

Figure 1-3. Running and editing programs

Creating a Program

Our first techBASIC program will turn on the accelerometer built into the iPhone or iPad; read the acceleration along the vertical axis, the horizontal axis, and through the screen of the device; record the time when the measurement was made; and print these values. It will then turn off the accelerometer to preserve battery life.

It sounds like a pretty sophisticated program, and it is. Here's the source code:

```
PRINT Sensors.accel
```

Let's see how it works. Sensors is the name of a class that is built into techBASIC. It's the class used to access all of the sensors that are built into the iPhone and iPad. One of the methods in the Sensors class is accel. It returns an array of four values: the current acceleration along the x-axis (vertically through the device), the y-axis (horizontally through the device), and the z-axis (perpendicular to the screen), and a timestamp indicating exactly when the reading was made.

PRINT is a statement in BASIC that prints any value, from numbers to strings to arrays. In this case, it prints all four values from the array returned by Sensors.accel.

Case Sensitivity

BASIC is not a case-sensitive language. The program shown will work just as well if you type:

```
print sensors.accel
```

The programs in this book follow a convention of showing all of the reserved words from the BASIC language in uppercase letters and capitalizing all class names. This is just a convention to make the programs easier for you to read. Use it or ignore it as you please.

It's time to enter the program and run it. If you are still in the *O'Reilly Books* folder, back up to the top folder level by tapping the Programs navigation button just above the list of programs.

From the program list, tap the *My Programs* folder. Tapping the name of a folder not only opens the folder, it also indicates which folder the program you're about to create will be placed in. Now tap the New button. You will see a dialog like Figure 1-4.

Figure 1-4. The New dialog

Enter Accel and tap OK. This creates a new, empty program. Enter the source code from our first program:

```
PRINT Sensors.accel
```

You should see something like what's shown in Figure 1-5.

Figure 1-5. The Accel program

If you're on an iPhone, you will need to dismiss the keyboard by tapping the Dismiss Keyboard button, then navigate back to the program list by tapping the Programs button. The Dismiss Keyboard button is the button at the top right that looks like a little keyboard with a triangle under it. You can dismiss the keyboard on the iPad, too, but it's optional.

Now tap the name of the program. On the iPad, you can also tap the Run button on the button bar. The screen will shift automatically to the console, which is the display that shows text input and output. Here's what I saw on my iPad when I ran the program:

```
-0.912109    -2.288818E-04  -0.394318     80395.372433
```

Acceleration is measured in units of gravities. These numbers show I was using my iPad in portrait mode, with the home button to the right. The acceleration along the x-axis is nearly –1, indicating that the left edge of the iPad was down. Apparently my desk is just about perfectly level, because acceleration along the y-axis was zero to three decimal places. The iPad was tipped back slightly on a folding stand, so the z acceleration was slightly negative. Figure 1-6 shows the coordinate system used by iPhone and iPad sensors. The time code doesn't tell you the actual time, just the number of seconds that have elapsed since some arbitrary time.

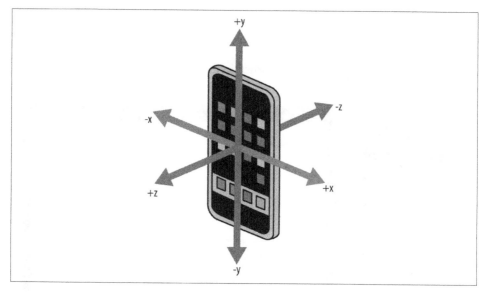

Figure 1-6. The coordinate system used by sensors on the iPhone and iPad—the coordinates stay fixed as the iPhone rotates, so –y always points to the home button

Run the program a few times, holding the iPhone or iPad in various orientations. You'll clearly see the sensor readings change.

The Accelerometer

While the simple program to read the accelerometer certainly does the job, we want something a bit more sophisticated. Our next program, seen in Figure 1-7, will read the accelerometer continuously, plotting the acceleration along each of the three axes on an oscilloscope-like display. Pressing the Record button records the data, writing it to a datafile that we can then read in other programs for subsequent processing. Of course, we may want to share the data or analyze it from another computer, so the Send button will send the most recent datafile to another device as an email attachment.

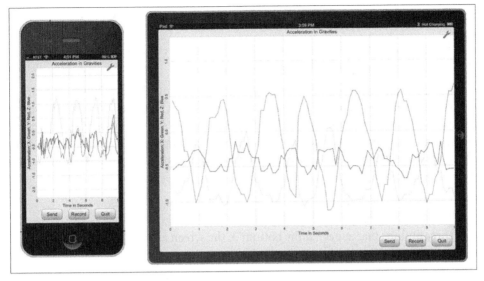

Figure 1-7. The Accelerometer app

Like all plots in techBASIC, this one can be resized and panned. Swipe up or down to see higher or lower values along the y-axis, or pinch vertically to see a wider range or to zoom in. techBASIC normally supports a few other features, like panning along the x-axis or tapping to see the value of a point on the plot, but these have been disabled, since they aren't necessary in this program.

Before we look at the source code, give the program a try to get familiar with it. Navigate to the *O'Reilly Books* folder and run the Accelerometer program. Be sure to record some data and email it to yourself. Try the swipe and pinch gestures to see how they work. Knowing how the program works will help as we dig through the source code.

 This program is a sample in techBASIC and techBASIC Sampler, so there is no need to type it in. Look for the program called Accelerometer in the *O'Reilly Books* folder.

Now that you have taken a moment to run the program and see what it does, let's dive in and see how it works. We'll break the program up into bite-sized pieces and explore how each piece works. Here's the first chunk, which you will see right at the top of the full program listing:

```
! Shows a running plot of the acceleration for the last 10
! seconds in 0.1-second intervals. Supports recording the
! values and emailing the results.

! Create the plots and arrays to hold the plot points.
```

```
DIM p as Plot, px as PlotPoint, py as PlotPoint, pz as PlotPoint
DIM ax(100, 2), ay(100, 2), az(100, 2)
```

The lines that start with an exclamation point are comments. They don't do anything; they exist solely so we can understand the program later.

The program shows the acceleration using three point plots that are shown on a single set of axes. The overall image is called a `Plot` in techBASIC, and a `PlotPoint` object handles each of the three point plots. These will be used in several places in the program, so they are defined in a `DIM` statement at the start of the program. The points that actually show up on the plot are stored in two-dimensional arrays; each array has 100 x, y pairs. These are defined in a second `DIM` statement.

```
! Create the controls.
DIM quit AS Button, record AS Button, send AS Button
```

The three buttons that appear at the bottom of the screen are defined next. Each will be a `Button` object.

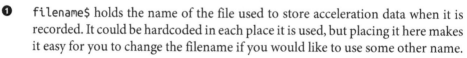

```
! Create and initialize the global tracking variables.
fileName$ = "tempdata.txt"  ❶
recording = 0  ❷
index = 1  ❸
```

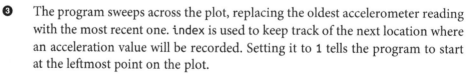

```
! Get and set an initial time for the accelerometer.
DIM t0 AS DOUBLE  ❹
a = Sensors.accel
t0 = a(4)
```

There are several variables that will be used in multiple places as the program runs, so these are initialized next:

❶ `filename$` holds the name of the file used to store acceleration data when it is recorded. It could be hardcoded in each place it is used, but placing it here makes it easy for you to change the filename if you would like to use some other name.

❷ `recording` is a flag used to indicate if the accelerometer data is being recorded or just displayed on the plot. The program will set this value to 1 whenever it is recording data.

❸ The program sweeps across the plot, replacing the oldest accelerometer reading with the most recent one. `index` is used to keep track of the next location where an acceleration value will be recorded. Setting it to 1 tells the program to start at the leftmost point on the plot.

❹ As you will see in a moment, the program will use a timer to tell when new accelerometer data is available, as well as when it was actually recorded by the accelerometer. `t0` is used to keep track of the time of the last accelerometer reading. It's actually initialized by taking an initial accelerometer reading, recording the time, and throwing away the acceleration values.

With all of the variables declared and initialized, the program calls a subroutine to set up the user interface.

```
! Create the user interface.
setUpGUI
```

If you glance at the complete source code for the program, you will see that the rest of the program is a collection of subroutines. On the iPad, you can see a list of the subroutines by tapping the Subs button shown in Figure 1-8. Setting up the user interface is the last thing the program does. The reason it doesn't just stop at this point is that it's an event-driven program. Two subroutines that handle events appear in the code. The program will continue to run, processing these events indefinitely, until we stop it. Since the program takes over the full screen, the obvious way to stop it is using the Quit button.

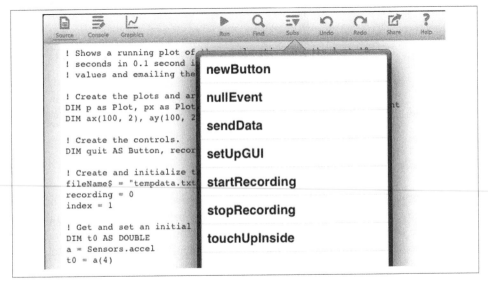

Figure 1-8. The subroutine list from the Subs button

There actually is another way to stop the program, though. Did you notice the small tool icon at the top right of the graphics screen? Tap this button and a menu of choices will show up, as shown in Figure 1-9.

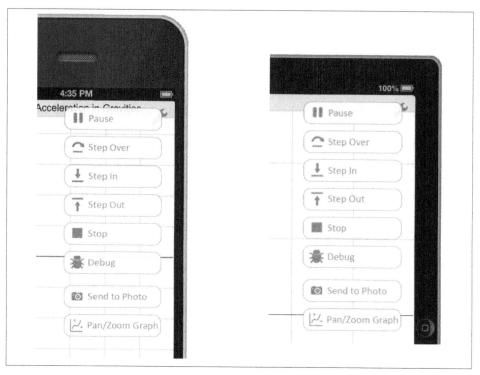

Figure 1-9. The tools icon and menu

One of the options is Stop. You can use this to stop a program you are working on, even if it has a bug that prevents the Quit button from working. There are several other options that give you access to the step-and-trace debugger, making it easier to find those pesky digital arthropods. See the techBASIC Quick Start Guides or Reference Manual for a quick but complete introduction to the debugger.

Getting the techBASIC Documentation

The techBASIC Reference Manual and Quick Start Guides are free downloads, available at the Byte Works website. (*http://www.byteworks.us/Byte_Works/Documenta tion.html*) They are PDF documents, so you can read them from iBooks on your iPad or iPhone.

There are lots of subroutines to look at. Let's start with the one that creates the user interface. From the iPhone, scroll down or use the Find button to search for the subroutine. From the iPad, tap the Subs button to get a list of subroutines, then tap setUp-GUI to jump right to that subroutine.

```
! Set up the user interface.

SUB setUpGUI
! Tell the accelerometer to update once every 0.05 seconds.
sensors.setAccelRate(0.05)
```

In theory, the accelerometer can take readings at a speed of up to about 100 samples per second. It won't be able to do it that quickly while we continuously update a plot, though. It also rarely records that fast even when a program is doing nothing but gathering accelerometer data. In any case, we're going to create a 10-second plot with 100 points, so we only need a new value about once every tenth of a second. It turns out that the accelerometer software in the operating system considers the data rate we specify to be a hint, not a commitment, and the data will come back at time intervals that don't exactly match the time we ask for. We'll ask for the data about twice as fast as we really need it, one point every 0.05 seconds, to make it more likely that we'll get a new value at least once every tenth of a second.

Why not just ask for it as fast as we can get it? Sensors use a fair amount of battery power, and they use more power at faster sampling rates. Only asking for the data we need saves the device's battery.

```
! Initialize the plot arrays.
FOR t = 1 TO 100
    ax(t, 1) = t/10.0
    ay(t, 1) = t/10.0
    az(t, 1) = t/10.0
NEXT
```

The three point plots each contain 100 x, y pairs used to locate the point that will be plotted on the screen. The y values will be set by the accelerometer as the program runs. The x values don't change—the program simply updates the y values as needed. These lines set the x values for each of the points to evenly distribute them across the plot. It uses values from 0.1 to 10.0, corresponding to the time value shown on the x-axis of the plot.

```
! Initialize the plot and show it.
p = Graphics.newPlot
```

The first step in creating the plot is to create the plot object that defines the axes. The variable p used to hold the Plot object was declared earlier in the program, but the plot doesn't exist until this line is executed.

```
p.setTitle("Acceleration in Gravities")
p.setXAxisLabel("Time in Seconds")
p.setYAxisLabel("Acceleration: X: Green, Y: Red, Z: Blue")
p.showGrid(1)
p.setGridColor(0.8, 0.8, 0.8)
```

Now that the plot exists, we can make it pretty, adding labels and creating a nice grid for a background.

```
p.setAllowedGestures($0042)
```

techBASIC supports a number of gestures to manipulate plots. For two-dimensional plots, swiping moves a plot around, pinching zooms in or out, and tapping shows the coordinates of a point on the plot. Since the horizontal axis is fixed and the points are constantly updating, most of these gestures are inappropriate in this program. This line turns all of them off except pinching and translation along the y-axis.

```
px = p.newPlot(ax)
px.setColor(0, 1, 0)
px.setPointColor(0, 1, 0)
```

These lines create the first of the three point plots. The first line creates the Plot Point object, saving it in the global variable px. The next two set the color of the connecting lines and points to green. This is the plot that will show acceleration along the x-axis, which is the horizontal axis as the iPhone or iPad is held with the home button down.

```
py = p.newPlot(ay)
py.setColor(1, 0, 0)
py.setPointColor(1, 0, 0)

pz = p.newPlot(az)
pz.setColor(0, 0, 1)
pz.setPointColor(0, 0, 1)
```

Here we do the same thing for the y- and x-axes, coloring the plots red and blue.

```
! Set the plot range and domain. This must be done
! after adding the first PlotPoint, since that also
! sets the range and domain.
p.setView(0, -2, 10, 2, 0)
```

Plots default to show the points in the first equation or set of points plotted. This line sets the view to show ±2G for 10 seconds.

```
! Show the graphics screen. Pass 1 as the parameter
! for full-screen mode.
system.showGraphics(1)
```

Now that almost everything is set up, the program switches to the graphics screen. Passing a 1 for the last parameter tells techBASIC to hide the development environment, using the full screen for the program.

```
! Lock the screen in the current orientation.
orientation = 1 << (System.orientation - 1)
System.setAllowedOrientations(orientation)
```

This is an accelerometer, and we expect anyone using it to bounce the iPhone about, turn it, and so forth. The iPhone normally responds to this by changing the screen orientation. There are two reasons we don't want that default behavior. The first is that it actually takes quite a bit of time for iOS to recreate the screen in a new orientation, and we don't want to lose any data while it's working. The second is that it's ugly.

These lines look to see what orientation the device is in when the app starts, and locks it into that orientation.

```
! Set the plot size.
p.setRect(0, 0, Graphics.width, Graphics.height - 47)
```

The default size for a plot is to fill the screen. We need room at the bottom for the buttons, so we manually set the size of the plot to leave 47 pixels at the bottom of the screen.

```
! Draw the background.
Graphics.setPixelGraphics(0)
Graphics.setColor(0.886, 0.886, 0.886)
Graphics.fillRect(0, 0, Graphics.width, Graphics.height)
```

That area at the bottom of the screen starts off white. This paints it the same light gray used for the background of the plot.

```
! Set up the user interface.
h = Graphics.height - 47
quit = newButton(Graphics.width - 82, h, "Quit")
record = newButton(Graphics.width - 174, h, "Record")
send = newButton(Graphics.width - 266, h, "Send")
```

Here we call a subroutine to create the three buttons at the bottom of the screen.

```
! If there is nothing to send, disable the Send button.
IF NOT EXISTS(fileName$) THEN
  send.setEnabled(0)
END IF
END SUB
```

We can't send the datafile in an email if it doesn't exist. If there isn't already a datafile from a previous run, `setEnabled(0)` disables the Send button so the users know it can't be used and don't think our program is broken when they tap the button and it does nothing.

Well, some of them will think it's broken, anyway. That's users for you. But at least we made an effort.

```
! Creates a new button with a gradient fill.
!
! Parameters:
!   x - Horizontal location.
!   y - Vertical location.
!   title - Name of the button.
!
! Returns: The new button.

FUNCTION newButton (x, y, title AS STRING) AS Button
DIM b AS Button
```

```
b = Graphics.newButton(x, y)
b.setTitle(title)
b.setBackgroundColor(1, 1, 1)
b.setGradientColor(0.6, 0.6, 0.6)
newButton = b
END FUNCTION
```

The subroutine that set up the user interface called this one to do the grunt work of creating a button. That's because there are several tasks that would have been repeated three times. It also makes the program easier to change. If you decide you want blue buttons, you just need to change the color once and all of the buttons will be updated.

The first part should look familiar, even though you haven't seen the program create a button yet. The button is declared with a DIM statement and created with a newButton call, and the title is set using setTitle.

The next two lines are the ones that give the nice gradient fill on the button, a small but professional touch that adds a lot to the overall impression people get when looking at a program. The lines vary the button color from white to a medium gray. There are lots of other options in techBASIC, like changing the angle of the gradient, but this simple gradient looks nice and is easy to create.

Finally, the newly created button is returned to the caller, where the setUpGUI subroutine stores it in the appropriate variable.

At this point we have a complete program, but if you were to run it, it would stop without doing anything interesting. It's time to tell techBASIC that we want some events.

```
! Handle a tap on one of the buttons.
!
! Parameters:
!    ctrl - The button that was tapped.
!    time - The time when the event occurred.

SUB touchUpInside (ctrl AS Button, time AS DOUBLE)
IF ctrl = quit THEN ❶
  stopRecording ❷
  STOP
ELSE IF ctrl = record THEN
  IF recording THEN ❸
    stopRecording
  ELSE
    startRecording
  END IF
ELSE IF ctrl = send THEN
  stopRecording ❹
  sendData
END IF
END SUB
```

Simply creating a subroutine with the name touchUpInside and this parameter list turns the program into an event-driven program. It will run until deliberately stopped, waiting for the user to tap on a button. When a button is finally tapped, this subroutine is called. It gets the button object, which we can use to figure out which of the three buttons was pressed, and the time when the button was tapped.

 The IF statement checks to see which button was tapped, comparing the parameter to the variables holding the three buttons. There are other ways to detect which button was pressed, but this one is easy to implement and understand.

❷ For the Quit button, the program stops any recording that might be in progress by calling stopRecording, then stops the program.

❸ The Record button actually does two different things. If the program is not recording the acceleration data, it starts; if the program is already recording acceleration data, it stops.

 Finally, the Send button stops any recording that might be in progress, then calls yet another subroutine to create an email and send the data.

All of the interesting stuff is happening in subroutines. Here's the first:

```
! Called when the program should start recording
! data, this subroutine changes the name of the
! recording button to Stop, opens the output file,
! and sets a flag indicating data should be
! recorded.

SUB startRecording
record.setTitle("Stop")
recording = 1
OPEN fileName$ FOR OUTPUT AS #1
END SUB
```

The record button does double duty, functioning both as a Record and a Stop button. This saves space, which is really at a premium on the iPhone, and also makes the program a little easier to write and perhaps a little easier to use by not showing and handling a lot of disabled buttons. The startRecording subroutine is called to start a recording. The user needs a way to stop it, too, so when the button is pressed one of the things the code does is change the name of the button to Stop.

We still haven't gotten to the code that actually reads the accelerometer, but when we do, it will need to know if it is supposed to record to the data file, or just draw what it reads on the plot. Setting recording to 1 is our signal to do both.

Finally, if we're going to write stuff to a file, we need to open the file. The OPEN command opens a file for output.

Files in BASIC

Files are one of those things that seem to be different in just about every programming language. If you're already good at a programming language other than BASIC, most of what you have seen up to this point should be pretty easy to read, even if the syntax is not quite what you are used to. The OPEN command may look odd, though.

BASIC files are referenced by a file number. There can be lots of them, but each unique file that is opened needs a unique number associated with that file. All file commands that manipulate the data in the file use this number. The number appears after a # character in all of the file-related commands. The open command in our subroutine is opening file number 1. Later, the program will write to the file with print statements that use the same number, like:

```
PRINT #1, a(1); ","; a(2); ","; a(3); ","; a(4)
```

and close the file with:

```
CLOSE #1
```

While you usually see the file number coded as a constant, it's okay to use a variable.

Files can be opened for OUTPUT, as is done here, or for INPUT. They can also be opened several other ways. While most programs in this book will read and write text data using standard BASIC PRINT and INPUT statements, it's also possible to get absolute control over a file, reading and writing binary data. See the techBASIC Reference Manual or any good book on BASIC for details.

```
! Called to stop recording data, this subroutine
! changes the name of the recording button to
! Recording, clears the recording flag and closes
! the output file.
!
! It is safe to call this subroutine even if
! nothing is being recorded.

SUB stopRecording
IF recording THEN
  record.setTitle("Record")
  CLOSE #1
  recording = 0
  send.setEnabled(1)
END IF
END SUB
```

The stopRecording subroutine undoes all of the actions taken by the startRecord
ing subroutine. The only finesse is that it checks to make sure a recording is in progress
before stopping it. That makes the program logic easier back in touchUpInside, as it

means the code to handle the Quit and Send buttons doesn't have to check to see if a recording is in progress. Instead, it can just call the subroutine, knowing that if a recording is in progress it will be stopped, but that it's safe to call the subroutine even if the program isn't recording the acceleration data.

```
! Send the last recorded data file to an email.

SUB sendData
DIM e AS eMail
e = System.newEMail
IF e.canSendMail THEN
  e.setSubject("Accelerometer data")
  e.setMessage("Accelerometer data")
  e.addAttachment(fileName$, "text/plain")
  e.send
ELSE
  button = Graphics.showAlert("Can't Send", _
      "Email cannot be sent from this device.")
END IF
END SUB
```

The code in touchUpInside that handles the Send button calls this subroutine to do the work. It defines and initializes the email object.

Not all iOS devices can actually send an email. Some may have the capability turned off, for example. The program starts by checking to see if it's possible to send an email, showing an alert if not.

If email is supported, it's a simple matter to attach the datafile, create a short subject and message, and send it. As you saw when you tried the program, pressing Send doesn't actually send the email, it just gets it ready and presents the mail message to the user. This gives the user a chance to address the email and send it manually.

It might seem nice if you could skip that step, addressing the email and sending it without user interaction. Apple blocks that action, though, and I think appropriately so; this ensures that rogue programs cannot collect and send information without the user's knowledge.

```
! Called when nothing else is happening, this
! subroutine checks to see if 0.1 seconds have
! elapsed since the last sensor reading. If so, a
! new one is recorded and displayed.
!
! Parameters:
!    time - The time when the event occurred.

SUB nullEvent (time AS DOUBLE)
a = Sensors.accel
```

The final subroutine is the one that actually collects and handles the acceleration data. This subroutine is called nullEvent. techBASIC calls it any time it is not busy doing something else, which in this program means any time except when a tap on a button is being handled. The subroutine starts by grabbing the most recent accelerometer value.

```
IF recording AND (t0 <> a(4)) THEN
   PRINT #1, a(1); ","; a(2); ","; a(3); ","; a(4)
END IF
```

Because this subroutine can be called pretty rapidly—more often than we've asked the accelerometer to collect acceleration information—we need to check to see if the time-stamp has changed since the last reading. If not, we've already handled this value and don't need to do so again. We also check to see if the program is supposed to save the acceleration data in the output file. If both conditions are met, the value from the accelerometer is written to the file.

The four numbers are separated by commas, and each reading is placed on a separate line. This is the so-called *comma-separated values*, or *CSV*, file format. It's extremely common, and a very easy format to write and read in BASIC. It's also a format that most spreadsheets and databases can read and write, so using it makes it easy to process the accelerometer data offline.

```
IF a(4) > t0 + 0.1 THEN
   WHILE a(4) > t0 + 0.1
     t0 = t0 + 0.1
     ax(index, 2) = a(1)
     ay(index, 2) = a(2)
     az(index, 2) = a(3)
     index = index + 1
     IF index > 100 THEN index = 1
   WEND
   px.setPoints(ax)
   py.setPoints(ay)
   pz.setPoints(az)
END IF
END SUB
```

The last section of the subroutine checks to see if at least one-tenth of a second has elapsed since the last value was placed in the plot. If so, it places the reading in the array holding the points to plot. Critically, it then loops to see if it needs to place the same point in again. This could happen if iOS was busy doing something else and didn't record a value for, say, a quarter of a second.

Finally, the arrays are passed to the various point plots using the setPoints method. This is where the plot actually gets updated.

As you've seen, almost all of the work is in creating and handling the user interface. There is a lot to creating a pleasant, usable user interface, so the program is a bit involved, but the effect is worth it.

We're also going to reuse most of the code for our next two programs!

CHAPTER 2

Accessing the Other Built-in Sensors

About This Chapter

Prerequisites
> The programs in this chapter are adaptations of the accelerometer from Chapter 1; read it first if this chapter seems confusing.

Equipment
> You will need an iPhone, iPod, or iPad running iOS 5 or later.

Software
> You will need a copy of techBASIC or techBASIC Sampler.

What You Will Learn
> This chapter shows how to access the magnetometer and gyroscope built into most iOS devices. You can use them for anything from direction finding to augmented reality.

Toward the end of the chapter you'll also learn a bit more about techBASIC, including another way to access the sensors that gives faster response times and how to use the techBASIC help system to find out more about these and other commands.

The chapter closes with a quick look at two other services. While not sensors in the traditional sense, they are accessed that way. One is GPS, and the other is heading, which uses the magnetometer and compass to find direction.

Writing our first GUI-based program to display sensor data from the iPhone and iPad in Chapter 1 was a bit involved. After all, you were learning a new programming environment as well as learning about the sensors. With that basic knowledge (pardon the pun), it's time to quickly expand what we can do. Among other things, we need to finish our tricorder!

The Gyroscope

Starting with the iPhone 4, all iPhones include a three-axis gyroscope. You might think a gyroscope is unnecessary, since the accelerometer can tell you the orientation of a device (as shown in Figure 2-1) and tracking it over time will tell you about changes in orientation, but it turns out the accelerometer really won't work well as a gyroscope substitute. One reason has to do with basic physics. The acceleration of gravity is not the only acceleration the accelerometer is exposed to; movement also causes acceleration. The other reason is practical. The accelerometer just can't detect rapid changes in orientation as well as a sensor like a gyroscope that is specifically designed for the task.

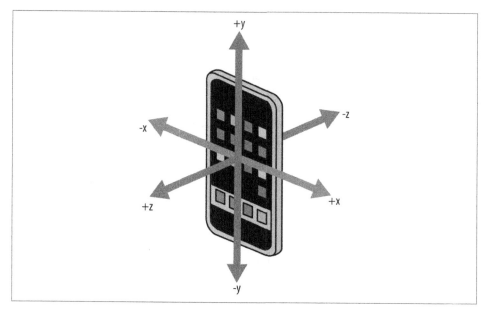

Figure 2-1. The orientation for the axes is the same for the accelerometer, gyroscope, and magnetometer

The gyroscope app we're going to build in this chapter (shown in Figure 2-2) looks a lot like the accelerometer app.

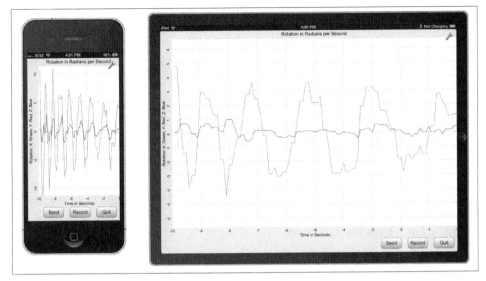

Figure 2-2. The Gyroscope app

The code is very similar, too. In fact, there are so few differences that we're not going to go through the program line by line. Let's look at the changes.

The first and most obvious change is that we're reading a different sensor. Instead of reading the accelerometer with a command like:

```
PRINT Sensors.accel
```

we read the gyroscope with the very similar command:

```
PRINT Sensors.gyro
```

There is another subtlety, though. Did you know that the gyroscope is only available in some models of the iPhone? That means it is very important to check to make sure it's there before we use it. This changes the program in two places. The first is right in the setup code, when we grab our initial value for the time:

```
! Get and set an initial time for the gyroscope.
DIM t0 AS DOUBLE
IF Sensors.gyroAvailable THEN
  WHILE t0 = 0
    r = Sensors.gyro
    t0 = r(4)
  WEND
END IF
```

The IF statement checks to see if the gyroscope is available before trying to read it. There is also a WHILE loop to make sure we get a nonzero time. That's because the gyroscope might return a zero time as it warms up.

The second change is at the end of `setUpGUI`:

```
! Make sure a gyroscope is available. If not, say
! so and stop the program.
IF NOT Sensors.gyroAvailable THEN
  msg$ = "This device does not have a gyroscope. "
  msg$ = msg$ & "The program will exit."
  button = Graphics.showAlert("No Gyro", msg$)
  STOP
END IF
```

This code checks to see if there is a gyroscope. If not, it puts up an alert to that effect and stops the program.

There is another minor update at the start of `setUpGUI`. The command to set the sensor rate must be changed to set the gyroscope sampling rate rather than the accelerometer sampling rate. Here's the new code:

```
SUB setUpGUI
! Tell the gyroscope to update once every 0.05 seconds.
Sensors.setGyroRate(0.05)
```

Running the program, you might notice that the running graph is handled in a slightly different way. The accelerometer app traced to the end, then wrapped around, updating from the beginning of the plot. The gyroscope app updates continuously, shifting the existing points to the left and placing the new one on the right. That requires two changes in the program. First, the `index` variable goes away, since it is no longer needed. The second change is in the `nullEvent` subroutine, where the code to update the plots changes from:

```
ax(index, 2) = a(1)
ay(index, 2) = a(2)
az(index, 2) = a(3)
index = index + 1
IF index > 100 THEN index = 1
```

to:

```
FOR i = 1 TO 99
  rx(i, 2) = rx(i + 1, 2)
  ry(i, 2) = ry(i + 1, 2)
  rz(i, 2) = rz(i + 1, 2)
NEXT
rx(100, 2) = r(1)
ry(100, 2) = r(2)
rz(100, 2) = r(3)
```

But doesn't it take more time to copy 297 values from one spot in an array to another than it does to simply update a single value and move on? Well, yes, but on a modern processor, it's really not going to affect things much. Copying the array values is very fast. Most performance issues will come from updating the graphics display. It's certainly worth considering issues like this, but the performance drop is not enough to see when the program is running.

Lots of Options

The accelerometer and gyroscope handle the plot update differently. Which do you like better? If you prefer one to the other, it's an easy matter to change the program so it behaves the way you prefer.

Since the program is now showing the current time at the right and going back 10 seconds into the past as you look farther left, the x-axis range has been changed from 0 to 10 to −10 to 0. The two changes needed to make this happen are in setUpGUI. The first is when the *x* values are initially assigned to the plot arrays:

```
! Initialize the plot arrays.
FOR t = 1 TO 100
   rx(t, 1) = t/10.0 - 10
   ry(t, 1) = t/10.0 - 10
   rz(t, 1) = t/10.0 - 10
NEXT
```

The second is where the visible range of the plot is initialized. Instead of showing 0 to 10 along the x-axis, the program shows −10 to 0:

```
! Set the plot range and domain. This must be done
! after adding the first PlotPoint, since that also
! sets the range and domain.
p.setView(-10, -10, 0, 10, 0)
```

If you have been comparing the code to the accelerometer app, you have probably already noticed the last, purely cosmetic change. Various variable names, comments, and labels were changed so they refer to the gyroscope rather than the accelerometer.

Here's the complete listing for the gyroscope app. It's also in techBASIC and techBASIC Sampler in the *O'Reilly Books* folder; look for the app called Gyroscope:

```
! Shows a running plot of rotation for the last 10 seconds
! in 0.1-second intervals. Supports recording the values
! and emailing the results.

! Create the plots and arrays to hold the plot points.
DIM p as Plot, px as PlotPoint, py as PlotPoint, pz as PlotPoint
DIM rx(100, 2), ry(100, 2), rz(100, 2)

! Create the controls.
DIM quit AS Button, record AS Button, send AS Button

! Create and initialize the global tracking variables.
fileName$ = "tempdata.txt"
```

```
recording = 0

! Get and set an initial time for the gyroscope.
DIM t0 AS DOUBLE
IF Sensors.gyroAvailable THEN
  WHILE t0 = 0
    r = Sensors.gyro
    t0 = r(4)
  WEND
END IF


! Create the user interface.
setUpGUI


! Creates a new button with a gradient fill.
!
! Parameters:
!    x - Horizontal location.
!    y - Vertical location.
!    title - Name of the button.
!
! Returns: The new button.

FUNCTION newButton (x, y, title AS STRING) AS Button
DIM b AS Button
b = Graphics.newButton(x, y)
b.setTitle(title)
b.setBackgroundColor(1, 1, 1)
b.setGradientColor(0.6, 0.6, 0.6)
newButton = b
END FUNCTION


! Called when nothing else is happening, this
! subroutine checks to see if 0.1 seconds have
! elapsed since the last sensor reading. If so, a
! new one is recorded and displayed.
!
! Parameters:
!    time - The time when the event occurred.

SUB nullEvent (time AS DOUBLE)
r = Sensors.gyro

IF recording AND (t0 <> r(4)) THEN
  PRINT #1, r(1); ","; r(2); ","; r(3); ","; r(4)
END IF

IF r(4) > t0 + 0.1 THEN
  WHILE r(4) > t0 + 0.1
    t0 = t0 + 0.1
```

```
    FOR i = 1 TO 99
       rx(i, 2) = rx(i + 1, 2)
       ry(i, 2) = ry(i + 1, 2)
       rz(i, 2) = rz(i + 1, 2)
    NEXT
    rx(100, 2) = r(1)
    ry(100, 2) = r(2)
    rz(100, 2) = r(3)
  WEND
  px.setPoints(rx)
  py.setPoints(ry)
  pz.setPoints(rz)
END IF
END SUB


! Send the last recorded data file to an email.

SUB sendData
DIM e AS eMail
e = System.newEMail
IF e.canSendMail THEN
   e.setSubject("Gyroscope data")
   e.setMessage("Gyroscope data")
   e.addAttachment(fileName$, "text/plain")
   e.send
ELSE
   button = Graphics.showAlert("Can't Send", _
        "Email cannot be sent from this device.")
END IF
END SUB


! Set up the user interface.

SUB setUpGUI
! Tell the gyroscope to update once every 0.05 seconds.
Sensors.setGyroRate(0.05)

! Initialize the plot arrays.
FOR t = 1 TO 100
  rx(t, 1) = t/10.0 - 10
  ry(t, 1) = t/10.0 - 10
  rz(t, 1) = t/10.0 - 10
NEXT

! Initialize the plot and show it.
p = Graphics.newPlot
p.setTitle("Rotation in Radians per Second")
p.setXAxisLabel("Time in Seconds")
p.setYAxisLabel("Rotation: X: Green, Y: Red, Z: Blue")
p.showGrid(1)
```

```
    p.setGridColor(0.8, 0.8, 0.8)
    p.setAllowedGestures($0042)

    px = p.newPlot(rx)
    px.setColor(0, 1, 0)
    px.setPointColor(0, 1, 0)

    py = p.newPlot(ry)
    py.setColor(1, 0, 0)
    py.setPointColor(1, 0, 0)

    pz = p.newPlot(rz)
    pz.setColor(0, 0, 1)
    pz.setPointColor(0, 0, 1)

    ! Set the plot range and domain. This must be done
    ! after adding the first PlotPoint, since that also
    ! sets the range and domain.
    p.setView(-10, -10, 0, 10, 0)

    ! Show the graphics screen. Pass 1 as the parameter
    ! for full-screen mode.
    system.showGraphics(1)

    ! Lock the screen in the current orientation.
    orientation = 1 << (System.orientation - 1)
    System.setAllowedOrientations(orientation)

    ! Set the plot size.
    p.setRect(0, 0, Graphics.width, Graphics.height - 47)

    ! Draw the background.
    Graphics.setPixelGraphics(0)
    Graphics.setColor(0.886, 0.886, 0.886)
    Graphics.fillRect(0, 0, Graphics.width, Graphics.height)

    ! Set up the user interface.
    h = Graphics.height - 47
    quit = newButton(Graphics.width - 82, h, "Quit")
    record = newButton(Graphics.width - 174, h, "Record")
    send = newButton(Graphics.width - 266, h, "Send")

    ! If there is nothing to send, disable the Send button.
    IF NOT EXISTS(fileName$) THEN
      send.setEnabled(0)
    END IF

    ! Make sure a gyroscope is available. If not, say
    ! so and stop the program.
    IF NOT Sensors.gyroAvailable THEN
      msg$ = "This device does not have a gyroscope. "
      msg$ = msg$ & "The program will exit."
```

```
    button = Graphics.showAlert("No Gyro", msg$)
    STOP
  END IF
END SUB


! Called when the program should start recording
! data, this subroutine changes the name of the
! recording button to Stop, opens the output file,
! and sets a flag indicating data should be
! recorded.

SUB startRecording
record.setTitle("Stop")
recording = 1
OPEN fileName$ FOR OUTPUT AS #1
END SUB


! Called to stop recording data, this subroutine
! changes the name of the recording button to
! Recording, clears the recording flag and closes
! the output file.
!
! It is safe to call this subroutine even if
! nothing is being recorded.

SUB stopRecording
IF recording THEN
  record.setTitle("Record")
  CLOSE #1
  recording = 0
  send.setEnabled(1)
END IF
END SUB


! Handle a tap on one of the buttons.
!
! Parameters:
!    ctrl - The button that was tapped.
!    time - The time when the event occurred.

SUB touchUpInside (ctrl AS Button, time AS DOUBLE)
IF ctrl = quit THEN
  stopRecording
  STOP
ELSE IF ctrl = record THEN
  IF recording THEN
    stopRecording
  ELSE
    startRecording
```

```
    END IF
  ELSE IF ctrl = send THEN
    stopRecording
    sendData
  END IF
END SUB
```

Radians or Degrees?

People in the physical sciences, engineers, and mathematicians are often familiar with radians, and are comfortable using the gyroscope's natural units for rotation of radians per second. But if your reaction to the units was, "radi-what?" it's easy enough to convert the program to show rotation in degrees per second instead of radians per second. Multiply the sensor readings by $180/\pi$ to convert the values reported by the gyroscope from radians per second to degrees per second. techBASIC has a handy function to do just that, called DEG. The easiest place to make the conversion is right after the sensor reading is taken. Add the FOR loop seen here to nullEvent:

```
SUB nullEvent (time AS DOUBLE)
r = Sensors.gyro

FOR i = 1 TO 3
  r(i) = DEG(r(i))
NEXT
```

The range of values will jump too, of course. Change the range of the plot in setUp GUI to ±500, like this:

```
! Set the plot range and domain. This must be done
! after adding the first PlotPoint, since that also
! sets the range and domain.
p.setView(-10, -500, 0, 500, 0)
```

Finally, change the plot title to reflect the new units. This line is also in setUpGUI:

```
! Initialize the plot and show it.
p = Graphics.newPlot
p.setTitle("Rotation in Degrees per Second")
```

The Magnetometer

Starting ((()))with the iPhone 3GS, all iPhones include a three-axis magnetometer. It's used mostly as a digital compass for map-based applications, but it can also measure surrounding magnetic fields. We'll put that ability to use to create a simple metal detector in Chapter 3, but we'll start with a very basic app here, shown in Figure 2-3. You can also use a magnetometer to find wires and other sources of current, or just to amuse yourself with refrigerator magnets.

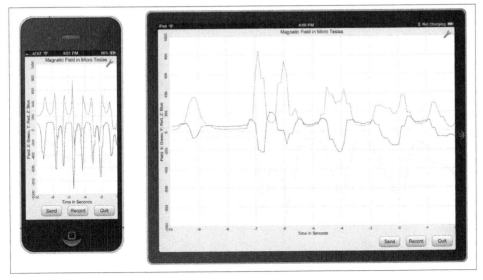

Figure 2-3. The Magnetometer app

The software for the magnetometer is almost identical to the software for the gyroscope. There is the obvious difference of reading a different sensor, but it looks pretty much like reading the accelerometer and gyroscope. Table 2-1 shows the equivalent commands.

Table 2-1. Sensor command equivalents

Accelerometer	Gyroscope	Magnetometer
Sensors.accel	Sensors.gyro	Sensors.mag
Sensors.accelAvailable	Sensors.gyroAvailable	Sensors.magAvailable
Sensors.setAccelRate	Sensors.setGyroRate	Sensors.setMagRate

There are also a number of obvious cosmetic changes, like changing variable names, comments, and strings so they make sense for the magnetometer.

The only change of any real substance is due to the huge variation in the strength of magnetic fields. The accelerometer tops out at ±2G for all iOS devices until the iPhone 5, which has a maximum range of ±8G. Setting the vertical scale to ±2G works well. The gyroscope rarely exceeds 10 radians per second, so setting the y-axis to ±10 worked well for our gyroscope app. The magnetometer, though, is sensitive enough to measure the strength of the Earth's magnetic field. That varies a bit, but is generally around 30–60 microteslas (30–60 µT). The sensor can also detect stronger magnetic fields; it tops out at 1 T, or 1,000,000 µT. Handling such a wide range of values well is a bit of a challenge. Here's the updated version of nullEvent that handles the range:

```
! Called when nothing else is happening, this
! subroutine checks to see if 0.1 seconds have
! elapsed since the last sensor reading. If so, a
! new one is recorded and displayed.
!
! Parameters:
!    time - The time when the event occurred.

SUB nullEvent (time AS DOUBLE)
! Get the new reading.
m = Sensors.mag

! If needed, write the value to the output file.
IF recording AND (t0 <> m(4)) THEN
  PRINT #1, m(1); ","; m(2); ","; m(3); ","; m(4)
END IF

! Update the plot.
IF m(4) > t0 + 0.1 THEN
  ! Update the arrays holding the points to plot.
  WHILE m(4) > t0 + 0.1
    t0 = t0 + 0.1
    FOR i = 1 TO 99
      mx(i, 2) = mx(i + 1, 2)
      my(i, 2) = my(i + 1, 2)
      mz(i, 2) = mz(i + 1, 2)
    NEXT
    mx(100, 2) = m(1)
    my(100, 2) = m(2)
    mz(100, 2) = m(3)
  WEND

  ! Adjust the function range based on the maximum observed value.
  max = 0
  FOR i = 1 TO 100
    IF ABS(mx(i, 2)) > max THEN max = ABS(mx(i, 2))
    IF ABS(my(i, 2)) > max THEN max = ABS(my(i, 2))
    IF ABS(mz(i, 2)) > max THEN max = ABS(mz(i, 2))
  NEXT
  range = 10^(INT(LOG(max)/LOG(10)) + 1)
  p.setView(-10, -range, 0, range, 0)

  ! Update the plots.
  px.setPoints(mx)
  py.setPoints(my)
  pz.setPoints(mz)
END IF
END SUB
```

The change is near the end, where the range is adjusted dynamically. The program scans the values to find the largest reading. It then uses a little mathematical trick, taking the integer of the base 10 logarithm of the largest value to find the correct power of 10 for the vertical range. Let's see how this works.

Let's start with a value that's about right for the Earth's magnetic field, 50 μT. In that case, LOG(50)/LOG(10) will be a tad under 1.7. Extract the integer part and add 1, and we have 2—the number of zeros we want for the range. Raise 10 to that power, and we get a range of ±100, perfect for displaying values from 10 to 100. Try a few other numbers to convince yourself that the formula will always return 100 for the range when max is 10 or greater and less than 100.

Pass a moderately strong magnet near the iPhone or iPad, and the field strength can jump to a few hundred μT. Try the math again, and you will find the range will be ±1,000 for values of max from 100 to 1,000.

It's kind of cool to see the program respond to the change in field strength by automatically adjusting the vertical scale of the plot, but there is a downside; manual adjustment no longer works. Well, it works, but the program resets the range as soon as the next measurement is taken. If this bothers you, remove the code that sets the range.

Here's the complete source for the magnetometer app. You can also find it in the *O'Reilly Books* folder in techBASIC and techBASIC Sampler. The program is called Magnetometer:

```
! Shows a running plot of the local magnetic field for
! the last 10 seconds in 0.1-second intervals. Supports
! recording the values and emailing the results.

! Create the plots and arrays to hold the plot points.
DIM p as Plot, px as PlotPoint, py as PlotPoint, pz as PlotPoint
DIM mx(100, 2), my(100, 2), mz(100, 2)

! Create the controls.
DIM quit AS Button, record AS Button, send AS Button

! Create and initialize the global tracking variables.
fileName$ = "tempdata.txt"
recording = 0

! Get and set an initial time for the magnetometer.
DIM t0 AS DOUBLE
IF Sensors.magAvailable THEN
  WHILE t0 = 0
    m = Sensors.mag
    t0 = m(4)
  WEND
END IF

! Create the user interface.
setUpGUI
```

```
! Creates a new button with a gradient fill.
!
! Parameters:
!    x - Horizontal location.
!    y - Vertical location.
!    title - Name of the button.
!
! Returns: The new button.

FUNCTION newButton (x, y, title AS STRING) AS Button
DIM b AS Button
b = Graphics.newButton(x, y)
b.setTitle(title)
b.setBackgroundColor(1, 1, 1)
b.setGradientColor(0.6, 0.6, 0.6)
newButton = b
END FUNCTION



! Called when nothing else is happening, this
! subroutine checks to see if 0.1 seconds have
! elapsed since the last sensor reading. If so, a
! new one is recorded and displayed.
!
! Parameters:
!    time - The time when the event occurred.

SUB nullEvent (time AS DOUBLE)
! Get the new reading.
m = Sensors.mag

! If needed, write the value to the output file.
IF recording AND (t0 <> m(4)) THEN
  PRINT #1, m(1); ","; m(2); ","; m(3); ","; m(4)
END IF

! Update the plot.
IF m(4) > t0 + 0.1 THEN
  ! Update the arrays holding the points to plot.
  WHILE m(4) > t0 + 0.1
    t0 = t0 + 0.1
    FOR i = 1 TO 99
      mx(i, 2) = mx(i + 1, 2)
      my(i, 2) = my(i + 1, 2)
      mz(i, 2) = mz(i + 1, 2)
    NEXT
    mx(100, 2) = m(1)
    my(100, 2) = m(2)
    mz(100, 2) = m(3)
  WEND
```

```
! Adjust the function range based on the maximum observed value.
max = 0
FOR i = 1 TO 100
  IF ABS(mx(i, 2)) > max THEN max = ABS(mx(i, 2))
  IF ABS(my(i, 2)) > max THEN max = ABS(my(i, 2))
  IF ABS(mz(i, 2)) > max THEN max = ABS(mz(i, 2))
NEXT
range = 10^(INT(LOG(max)/LOG(10)) + 1)
p.setView(-10, -range, 0, range, 0)

! Update the plots.
px.setPoints(mx)
py.setPoints(my)
pz.setPoints(mz)
END IF
END SUB


! Send the last recorded data file to an email.

SUB sendData
DIM e AS eMail
e = System.newEMail
IF e.canSendMail THEN
  e.setSubject("Magnetometer data")
  e.setMessage("Magnetometer data")
  e.addAttachment(fileName$, "text/plain")
  e.send
ELSE
  button = Graphics.showAlert("Can't Send", _
      "Email cannot be sent from this device.")
END IF
END SUB


! Set up the user interface.

SUB setUpGUI
! Tell the magnetometer to update once every 0.05 seconds.
Sensors.setMagRate(0.05)

! Initialize the plot arrays.
FOR t = 1 TO 100
  mx(t, 1) = t/10.0 - 10
  my(t, 1) = t/10.0 - 10
  mz(t, 1) = t/10.0 - 10
NEXT

! Initialize the plot and show it.
p = Graphics.newPlot
p.setTitle("Magnetic Field in Micro Teslas")
p.setXAxisLabel("Time in Seconds")
```

```
p.setYAxisLabel("Field: X: Green, Y: Red, Z: Blue")
p.showGrid(1)
p.setGridColor(0.8, 0.8, 0.8)
p.setAllowedGestures($0042)

px = p.newPlot(mx)
px.setColor(0, 1, 0)
px.setPointColor(0, 1, 0)

py = p.newPlot(my)
py.setColor(1, 0, 0)
py.setPointColor(1, 0, 0)

pz = p.newPlot(mz)
pz.setColor(0, 0, 1)
pz.setPointColor(0, 0, 1)

! Set the plot range and domain. This must be done
! after adding the first PlotPoint, since that also
! sets the range and domain.
p.setView(-10, -10, 0, 10, 0)

! Show the graphics screen. Pass 1 as the parameter
! for full-screen mode.
system.showGraphics(1)

! Lock the screen in the current orientation.
orientation = 1 << (System.orientation - 1)
System.setAllowedOrientations(orientation)

! Set the plot size.
p.setRect(0, 0, Graphics.width, Graphics.height - 47)

! Draw the background.
Graphics.setPixelGraphics(0)
Graphics.setColor(0.886, 0.886, 0.886)
Graphics.fillRect(0, 0, Graphics.width, Graphics.height)

! Set up the user interface.
h = Graphics.height - 47
quit = newButton(Graphics.width - 82, h, "Quit")
record = newButton(Graphics.width - 174, h, "Record")
send = newButton(Graphics.width - 266, h, "Send")

! If there is nothing to send, disable the Send button.
IF NOT EXISTS(fileName$) THEN
  send.setEnabled(0)
END IF

! Make sure a magnetometer is available. If not, say
! so and stop the program.
IF NOT Sensors.magAvailable THEN
```

```
    msg$ = "This device does not have a magnetometer. "
    msg$ = msg$ & "The program will exit."
    button = Graphics.showAlert("No Magnetometer", msg$)
    STOP
  END IF
END SUB


! Called when the program should start recording
! data, this subroutine changes the name of the
! recording button to Stop, opens the output file,
! and sets a flag indicating data should be
! recorded.

SUB startRecording
record.setTitle("Stop")
recording = 1
OPEN fileName$ FOR OUTPUT AS #1
END SUB


! Called to stop recording data, this subroutine
! changes the name of the recording button to
! Recording, clears the recording flag and closes
! the output file.
!
! It is safe to call this subroutine even if
! nothing is being recorded.

SUB stopRecording
IF recording THEN
  record.setTitle("Record")
  CLOSE #1
  recording = 0
  send.setEnabled(1)
END IF
END SUB


! Handle a tap on one of the buttons.
!
! Parameters:
!     ctrl - The button that was tapped.
!     time - The time when the event occurred.

SUB touchUpInside (ctrl AS Button, time AS DOUBLE)
IF ctrl = quit THEN
  stopRecording
  STOP
ELSE IF ctrl = record THEN
  IF recording THEN
    stopRecording
```

```
    ELSE
      startRecording
    END IF
  ELSE IF ctrl = send THEN
    stopRecording
    sendData
  END IF
END SUB
```

Faster Sensor Response

The three programs presented so far are fun, useful ways to see the raw sensor values for the accelerometer, gyroscope, and magnetometer, but they do have a drawback. Because so much time is spent updating a graphical environment, the sample rate is limited. It's possible to ditch the graphics and loop over a call to read the sensors, and that's considerably better, but even that method could lose data. There is another way to sample the sensors in techBASIC that gives considerably faster response rates—as fast as the iPhone supports—but the price is heavy. You make the call, and the program stops responding until all of the sensor data has been collected. Let's explore this command, and discuss other resources for learning more about techBASIC at the same time.

techBASIC has a built-in help system with technical information about every statement, function, and class in the program. From the iPad, tap the Source button to look at a program's source code, then tap the Help button on the toolbar. From the iPhone, you need to be editing a program to see the Help button. When you tap this button, what you see is a list of the features by category, as shown in Figure 2-4.

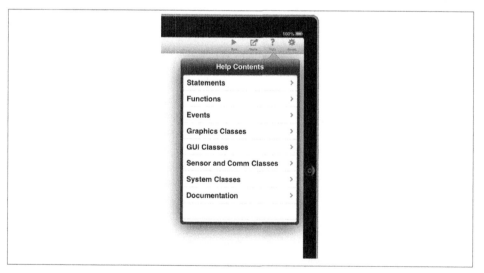

Figure 2-4. The built-in help system

Program statements, like the PRINT command from the very first short program, are in the Statements category. Functions, like the DEG function that converts radians to degrees, are in the Functions category. All of the event subroutines are in Events; these are the special subroutines like nullEvent and touchUpInside that techBASIC calls in event-driven programs. The rest of the categories are groups of predefined classes.

We want to explore the Sensors class, described in the Sensor and Comm Classes category shown in Figure 2-5. Tap that name. Next, you'll see the various classes used for accessing internal and external sensors. We'll be using a lot of these in the book, so you will get very familiar with this section!

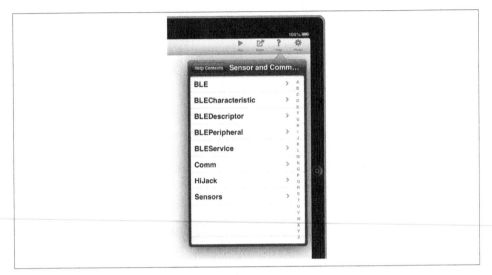

Figure 2-5. The Sensor and Comm Classes category

The last class listed is the Sensors class that we've used so much in this chapter. Tap that name. All class descriptions start with an overview that gives a broad description of the class, as shown in Figure 2-6. You'll also see some methods that should be very familiar to you. We're looking for a new one, though. Scroll down and tap on the sample method.

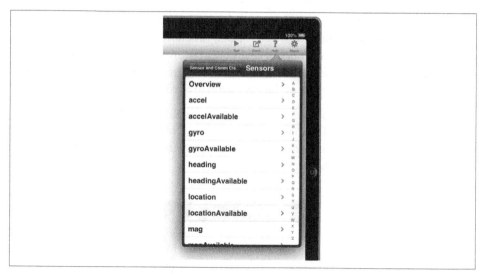

Figure 2-6. The Sensors class

Figure 2-7 shows a complete description of the sample method. It can read all three sensors at once, and can read them as fast as the operating system will allow. The description gives all of the technical details about the call; everything, in fact, that you need to write a program using the call.

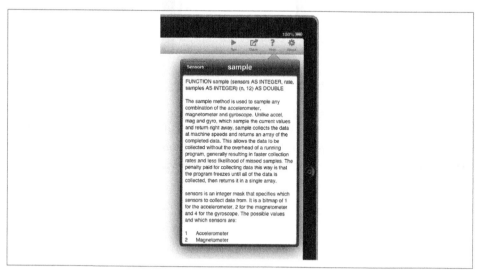

Figure 2-7. The sample method

There is another resource that takes this even farther. The help system is designed for quick reference and small size. It does not have many code samples, and there are no illustrations. There is a comprehensive techBASIC Reference Manual, though, that offers both.

 The techBASIC Reference Manual is a free download, available at the Byte Works website (*http://www.byteworks.us/Byte_Works/Documen tation.html*). It's a PDF document, so you can read it from iBooks on your iPad or iPhone.

Flipping to the section that describes the `sample` method, you'll find this code snippet. Like most of the code snippets in the manual, it's a complete, functioning program:

```
! Collect sensor samples for 5 seconds.
samples = 250
samp = Sensors.sample(7, 0.02, samples)

! Find the average and maximum values for each
! sensor.
FOR i = 1 TO samples
  ax = ax + samp(i, 1)
  ay = ay + samp(i, 2)
  az = az + samp(i, 3)

  IF ABS(max) < ABS(samp(i, 1)) THEN max = samp(i, 1)
  IF ABS(may) < ABS(samp(i, 2)) THEN may = samp(i, 2)
  IF ABS(maz) < ABS(samp(i, 3)) THEN maz = samp(i, 3)

  gx = gx + samp(i, 5)
  gy = gy + samp(i, 6)
  gz = gz + samp(i, 7)

  IF ABS(mgx) < ABS(samp(i, 5)) THEN mgx = samp(i, 5)
  IF ABS(mgy) < ABS(samp(i, 6)) THEN mgy = samp(i, 6)
  IF ABS(mgz) < ABS(samp(i, 7)) THEN mgz = samp(i, 7)

  mx = mx + samp(i, 9)
  my = my + samp(i, 10)
  mz = mz + samp(i, 11)

  IF ABS(mmx) < ABS(samp(i, 9)) THEN mmx = samp(i, 9)
  IF ABS(mmy) < ABS(samp(i, 10)) THEN mmy = samp(i, 10)
  IF ABS(mmz) < ABS(samp(i, 11)) THEN mmz = samp(i, 11)
NEXT

PRINT USING "Max acceleration: ##.##, ##.##, ##.##"; max, may, maz
PRINT USING "Max rotation: ###.##, ###.##, ###.##"; mgx, mgy, mgz
PRINT USING "Max magnetic field: ###.##, ###.##, ###.##"; mmx, mmy, mmz
```

```
PRINT USING "Average acceleration: ##.##, ##.##, ##.##"; _
         ax/samples, ay/samples, az/samples
PRINT USING "Average rotation: ###.##, ###.##, ###.##"; _
         gx/samples, gy/samples, gz/samples
PRINT USING "Average magnetic field: ###.##, ###.##, ###.##"; _
         mx/samples, my/samples, mz/samples
```

While it's not fancy, this complete program samples all three sensors at a rate of 50 samples per second for 5 seconds. It then prints the maximum and average values recorded.

This program is not one of the techBASIC samples. So how do you get it into techBASIC? Unfortunately, Apple is a tad overprotective of the iPhone here. Silly as it may be, there are only two ways to get the program to techBASIC. The first is to type it in. The second is to email it to yourself, then copy it from the email and paste it into a new, empty program. Still, even using an email isn't all that bad.

Try that and see how you like the program. You know enough now to add some flourish if you like.

Heading

You may have noticed two other services in the Sensors class as you opened the description of the sample method. The first is heading, which is really a precomputed composite of information gleaned from the magnetometer, accelerometer, and GPS. The task of finding direction from this information is not trivial, as we'll see in the next chapter. For now, though, it's enough to know that the heading is pretty easy to find. As with the three sensors we've used so far, it's just a matter of reading the value, perhaps after using headingAvailable to make sure the service exists on the device being used. There is a catch, though. The first call to heading starts the service, but it takes a while to get going. Rather than lock up the program while it waits, techBASIC returns the most recently recorded value. You should take a few readings, not just one:

```
FOR I = 1 TO 10
  PRINT Sensors.heading
  System.wait(0.5)
NEXT
```

This program reads the heading 10 times, waiting for a half second between readings. When I ran it on my iPad, here's what I got (with some blank lines removed):

```
0              0            -1          22182.984556
90.786407      81.723602    -1          22183.473079
92.786407      83.723602    -1          22183.965387
38.786407      29.723602    40          22184.478982
34.786407      25.723602    40          22184.987621
34.786407      25.723602    40          22185.500116
33.786407      24.723602    40          22185.989068
33.786407      24.723602    40          22186.506593
```

```
33.786407      24.723602      40      22186.99292
33.786407      24.723602      40      22187.501888
```

As you can see, the first reading is garbage.

Starting with the second reading, the first two numbers are the true heading and magnetic heading, just like they would appear on a compass. It appears that my iPad was pointed roughly to the east, then was turned to point a bit east of north before the fourth reading.

Actually, though, I didn't turn it at all—which points out the need for the third value returned. It's the accuracy of the reading, which was –1 for the first three readings. That means these readings are not accurate at all, and reminds us that it takes a while for the heading to return valid values. Even after the heading settles down, the accuracy is not that good; iOS is only claiming the heading is good to within 40 degrees.

The last value is the time when the reading was taken.

Location

The last service from the Sensors class is the location. Is that a new internal sensor, or a report from the external sensors made up of the GPS satellite system? I think of it as a combination. In any case, it's used pretty much like the heading and, like the heading, takes some time to warm up. Here's a short sample that shows how to read the location. It's copied directly from the snippet illustrating the location command in the techBASIC Reference Manual:

```
location = sensors.location(30)
PRINT USING "Latitude      :  ####.###"; location(1)
PRINT USING "Longitude     :  ####.###"; location(2)
PRINT USING "Altitude      : #####.##"; location(3)
PRINT USING "Horiz. Error:  ####.###"; location(4)
PRINT USING "Vert. Error :  ####.###"; location(5)
PRINT USING "Speed         :  ####.###"; location(6)
PRINT USING "Direction     :  ####"; location(7)
PRINT        "Time stamp  : "; location(8)
```

Running it from my desk, here's what I got with a few digits of precision removed from the location—it's pretty accurate!

```
Latitude     :      35.xxx
Longitude    :    -106.xxx
Altitude     :    1514.xx
Horiz. Error:      65.000
Vert. Error  :      44.790
Speed        :      -1.000
Direction    :      -1
Time stamp  : 24355.097656
```

The first five values are pretty self-explanatory, but keep in mind that the units for distance are meters.

Speed and direction are calculated from a series of locations. They are –1 if the speed or direction can't be calculated, as was the case for my iPad sitting still on my desk. The direction is not the compass heading, but the actual direction of travel.

Last, of course, is the timestamp when the measurement was made.

Your Own Tricorder

So now you have your own tricorder. It can display, record, and email acceleration, rotation, and magnetic field strength data. You've also learned how to access the heading, location, direction of travel, and speed of travel using a few more calls.

Chapter 3 will put this knowledge to real-world work, developing a metal detector based on the magnetometer. Much later, in Chapter 7, we'll return to accelerometers, although that accelerometer will be on a remote device. That's when we'll investigate how to find speed and distance traveled using acceleration data, when we use the accelerometer to find the altitude of a model rocket.

Creating a Metal Detector

About This Chapter

Prerequisites
Read Chapter 1 and the end of Chapter 2 (the section on the techBASIC help system) if you need some help with techBASIC. If you are already comfortable with tech-BASIC, jump right in.

Equipment
You will need an iPhone, iPod, or iPad running iOS 5 or later.

Software
You will need a copy of techBASIC or techBASIC Sampler.

What You Will Learn
This is a project chapter, using knowledge from Chapter 1 and Chapter 2 to develop a metal detector. There's also a lot of background material explaining why the metal detector works and, in some cases, why it doesn't. It always helps to know the limitations of an instrument!

The magnetometer in the iPhone is an amazing instrument that gives you the magnetic field strength along three axes in microteslas (μT). Apple software uses this instrument to find the direction in which the iPhone is pointed. This chapter looks at how the magnetometer works, shows you how to access the data from techBASIC, and illustrates using the magnetometer to develop a metal detector.

The iPhone/iPad Magnetometer

If you're like me, just using a metal detector is not all that much fun. I want to know how it ticks. That's what the first few parts of this chapter are about. If you don't care, skip to "Using the iPhone or iPad as a Metal Detector" on page 48, where you'll find a description of how to use the metal detector app.

Starting with the iPhone 3GS, all iPhones have a built-in magnetometer. All iPads have one, but the iPod Touch does not. But what is a magnetometer, really? Obviously it measures magnetic fields, but how does it do that?

You might recall from basic physics that electric currents and magnetic fields are related. In fact, current flowing through a wire causes a magnetic field around the wire. Wrap an insulated wire a few dozen times around a large iron nail and apply a current from a battery. The magnetic fields build up from each loop of wire, turning the nail into a magnet. Cut the current, and the nail is no longer a magnet. The same thing works in reverse, too. If you pass the nail through a magnetic field, it will generate a current in the wire. Magnetic fields move the electrons. Either the electrons or the magnetic field need to be moving, though. If they are both stationary, the effect stabilizes.

Now imagine running a current through a wire and applying a magnetic field to the wire at the same time. The moving electrons will shift from an even distribution in the wire, moving very slightly to one side. The stronger the magnetic field, the more they will move. If you can measure this shift in the current, you can tell if there is a magnetic field present, and how strong it is. This is called the *Hall effect*, and it's how the AK8973 chip detects a magnetic field. (For more information on the Hall effect, search for it on Wikipedia (*http://en.wikipedia.org/wiki/Hall_effect*).) The AK8973 chip is the one used in the iPhone 3GS. This chip, or a close cousin, is used in the other iPhone and iPad models. For an in-depth look at the AK8973, go to Alldatasheet.com, where you can download the specifications for the chip (*http://www.alldatasheet.com/datasheet-pdf/pdf/219477/AKM/AK8973.html*).

The strength of a magnetic field is generally measured in *teslas* (T), an SI measurement unit named after the famous Nikola Tesla. The units for a tesla are:

$$1T = 1\frac{V \cdot s}{m^2}$$

where V is a volt, s is a second, and m is a meter. The iPhone can measure magnetic fields up to about 1T with a precision of about 8 μT (8 microteslas).

The Earth's Magnetic Field

Of course, the reason Apple put a magnetometer in the iPhone was to tell which direction it was pointed in, which, along with the GPS, allows for some really cool navigation apps. You can find magnetic north by measuring the direction of the Earth's magnetic field. But finding true north—the direction of the physical north pole—is not quite so simple.

The first problem you have to deal with is that the Earth's magnetic field isn't exactly parallel with the north and south poles. Even worse, it has local variations from place to place. It turns out that to find true north you need to know the direction of magnetic north and the location you are at! Figure 3-1 provides a map that shows the declination of the Earth's magnetic field, which is the difference between magnetic north and true north. The zero lines, like the one running almost vertically through North and South America, are the lucky locations where magnetic north and true north are aligned. For positive declination, like the area covering most of the Pacific Ocean, the magnetic field is a little east of true north; for negative declination, like the lines covering the Atlantic Ocean, it is a little to the west.

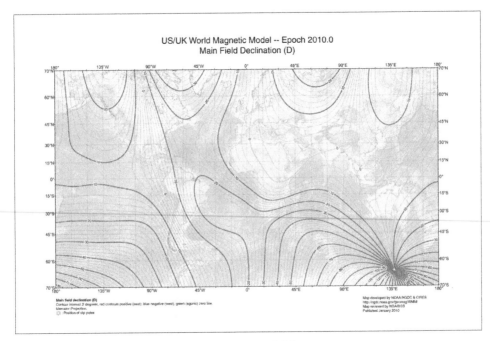

Figure 3-1. Declination of the Earth's magnetic field

Problem solved, right? We can find our location with the GPS, and magnetic north with the AK8973 chip. A quick table lookup, perhaps with some interpolation if we want to get fancy, and we have the direction of true north. Well, yes—but the problem is that the AK8973 chip won't really give us magnetic north. It's trapped inside a metal box, with currents flowing all around it and a speaker magnet nearby. All of those things affect the chip, altering the direction it reports as magnetic north. Fortunately, the currents, case, and speaker magnets don't move around, so we can calculate their effects on the field once and adjust for them. This is called the *magnetic deviation*, and it's an even bigger problem in places like airplanes, which have even bigger metal boxes surrounding the compass. We're going to finesse this problem for the metal detector, but it's one you'll have to deal with for other magnetometer applications—more on that later.

Using the iPhone or iPad as a Metal Detector

Earlier we talked about the fact that currents can create magnetic fields, and magnetic fields can create currents. Metal also affects a magnetic field. A metal is a material that has at least one electron that is loosely bound to the parent atom, making it relatively easy to move the electrons from one atom to another. All those mobile electrons bend the field a bit, changing its strength and direction. And that's how we're going to turn your iPhone into a metal detector.

First, let's do a quick experiment to find out if all of this is possible. The strength of the Earth's magnetic field varies from place to place. Here in Albuquerque, it's about 50μT. Can a small piece of metal distort such a small magnetic field enough to be useful? I ran the Magnetometer sample from Chapter 2 and passed my iPhone about six inches over my aluminum-topped keyboard to find out. Figure 3-2 shows what I saw.

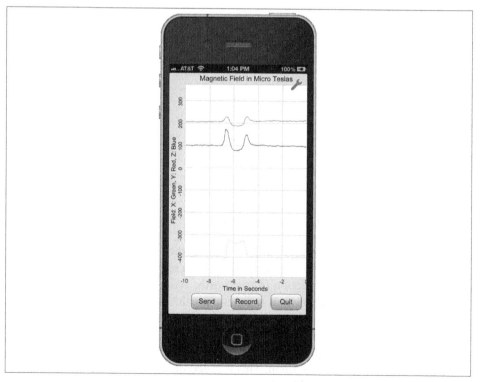

Figure 3-2. Magnetic field as the iPhone passes over metal

OK, that's promising. The first red and blue spike and the leading edge of the green spike occurred just as I hit the edge of the keyboard. However, this is a case of too much information. I don't really want the magnetic field in three directions; turning the iPhone as I move it could give me a false positive. Figure 3-3 shows what I get just from turning the iPhone.

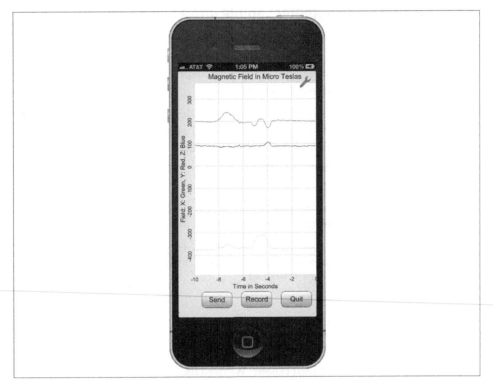

Figure 3-3. Variation in the magnetic field from turning the iPhone

It may not be obvious yet, but we are seeing two effects on the magnetometer here. The first, and biggest, comes from simply twisting the detector in the Earth's magnetic field. As you can see, for the most part, I was twisting the iPhone about the z-axis, changing the field strength in the y and z directions, and these more or less counterbalance one another. That's what we would expect. There is another effect that is hard to see here, which we will return to in a moment. In any case, it's clear that our magnetometer would be a lot more useful if we could see the overall field strength instead of the strength along each axis. We can find the overall field strength by resolving the three vectors into a single magnitude, as shown here:

$$m = \sqrt{x \cdot x + y \cdot y + z \cdot z}$$

or, in BASIC:

```
m = SQR(x*x + y*y + z*z)
```

where x, y, and z are the magnetic field strengths along each axis, and m is the overall field strength.

Converting the Magnetometer Sample into a Metal Detector

The complete metal detector source code is available in techBASIC and techBASIC Sampler in the *O'Reilly Books* folder. It's interesting how easy it is to take the existing Magnetometer sample and turn it into another useful program, though. The rest of this section looks at what changes were needed.

The goal was to modify the magnetometer sample to show the overall magnitude of the magnetic field instead of the components along each axis. There are different ways to do this. I deleted the two extra graphs, changed a couple of labels and comments, and plotted the magnitude of the total magnetic field instead of the magnitude in the x direction.

The first change is at the top of the program, where we only need one PlotPoint object and one array for points. The magnetometer app has these lines:

```
! Create the plots and arrays to hold the plot points.
DIM p as Plot, px as PlotPoint, py as PlotPoint, pz as PlotPoint
DIM mx(100, 2), my(100, 2), mz(100, 2)
```

These change to:

```
! Create the plots and arrays to hold the plot points.
DIM p as Plot, pm as PlotPoint
DIM mm(100, 2)
```

Most of the other changes are in nullEvent. We need the overall magnetic field strength, so that calculation is added right after the magnetic field strength is collected:

```
SUB nullEvent (time AS DOUBLE)
! Get the new reading.
m = Sensors.mag

! Calculate the overall magnetic field strength.
fieldStrength = SQR(m(1)*m(1) + m(2)*m(2) + m(3)*m(3))
```

Right after this is the section of code that writes information to a file if one is open for recording. There are two reasonable choices here. One is to save just the overall magnetic field strength, since that's what the metal detector app is showing. More data is often useful, though, so the app actually saves both the raw data and the overall field strength:

```
! If needed, write the value to the output file.
IF recording AND (t0 <> m(4)) THEN
```

```
    PRINT #1, m(1); ","; m(2); ","; m(3); ","; m(4); ","; fieldStrength
  END IF
```

The remaining part of nullEvent updates the plot. This code is simpler because it only has one plot to update, not three, but is otherwise the same:

```
! Update the plot.
IF m(4) > t0 + 0.1 THEN
  ! Update the arrays holding the points to plot.
  WHILE m(4) > t0 + 0.1
    t0 = t0 + 0.1
    FOR i = 1 TO 99
      mm(i, 2) = mm(i + 1, 2)
    NEXT
    mm(100, 2) = fieldStrength
  WEND

  ! Adjust the function range based on the maximum observed value.
  max = 0
  FOR i = 1 TO 100
    IF ABS(mm(i, 2)) > max THEN max = ABS(mm(i, 2))
  NEXT
  range = 10^(INT(LOG(max)/LOG(10)) + 1)
  p.setView(-10, -range, 0, range, 0)

  ! Update the plots.
  pm.setPoints(mm)
END IF
END SUB
```

There are some similar changes at the start of setUpGUI, where the plot and plot array are initialized. Again, it's essentially the magnetometer app, just simpler because of the reduced number of plots. There are also a couple of cosmetic changes to the labels on the plot:

```
SUB setUpGUI
! Tell the magnetometer to update once every 0.05 seconds.
Sensors.setMagRate(0.05)

! Initialize the plot arrays.
FOR t = 1 TO 100
  mm(t, 1) = t/10.0 - 10
NEXT

! Initialize the plot and show it.
p = Graphics.newPlot
p.setTitle("Magnetic Field Strength in Micro Teslas")
p.setXAxisLabel("Time in Seconds")
p.setYAxisLabel("Field Strength")
p.showGrid(1)
p.setGridColor(0.8, 0.8, 0.8)
p.setAllowedGestures($0042)
```

```
pm = p.newPlot(mm)
pm.setColor(1, 0, 0)
pm.setPointColor(1, 0, 0)
```

Pretty cool, right? Those are some fairly minor changes to a program, and suddenly it's doing something different.

Using the Metal Detector

Run the Metal Detector sample from the *O'Reilly Books* folder in techBASIC. Holding your iPhone away from any metallic objects, turn it slightly. This way, you ensure that only the Earth's magnetic field has an effect. Figure 3-4 shows what I saw after using a swipe and pinch to zoom in on the interesting part of the plot.

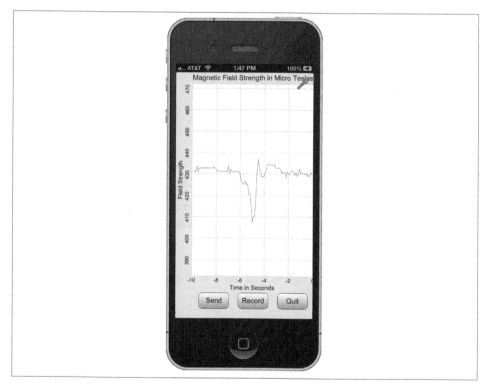

Figure 3-4. Magnetic field changes from twisting the iPhone

You might expect the magnetic field to be constant, since the magnitude of the Earth's magnetic field is not changing as the iPhone rotates. That's not what the data shows, though. Why? Well, consider that the magnetometer is, in fact, a metal detector and an electric field detector that is sensing changes in the Earth's magnetic field as metal and current warp that field. As we mentioned earlier, the iPhone itself has a lot of metal in it, probably has a magnet in the speaker, and definitely has electric currents. The metal detector is detecting the iPhone! Fortunately, we can largely eliminate this effect by holding the iPhone in a constant orientation as we pass it over the suspected metal. As long as we don't twist the iPhone as we move it, the distortion of the Earth's magnetic field from the interaction of the iPhone does not change.

Passing the metal detector over my keyboard without much rotation gives the result shown in Figure 3-5. The spike occurred just as the iPhone passed over the edge of the keyboard. Speed is important, too. Move the iPhone too fast or too slow, and the size of the spike is reduced.

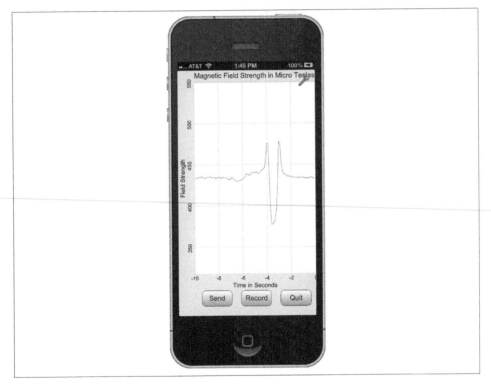

Figure 3-5. Detecting metal with the Metal Detector app

If you are having difficulty seeing the change, especially if the plot looks like a straight line, remember that you can increase the vertical scale by placing two fingers on the screen and spreading them out in a vertical direction.

So there you have it: your iPhone is a metal detector! You can experiment with various metallic objects to get an idea of how sensitive it is and how fast to move the iPhone for the best results.

Finding Out More

Wikipedia contains several useful articles that give background and details about some of the topics covered in this chapter:

- Hall effect (*http://en.wikipedia.org/wiki/Hall_effect*)
- Earth's magnetic field (*http://en.wikipedia.org/wiki/Earth's_magnetic_field*)
- Magnetic declination (*http://en.wikipedia.org/wiki/Magnetic_declination*)
- Magnetic deviation (*http://en.wikipedia.org/wiki/Magnetic_deviation*)
- More on metal detectors (*http://en.wikipedia.org/wiki/Metal_detectors*)

You can also check out the datasheet "AK8973 – 3-axis Electronic Compass" (*http://www.alldatasheet.com/datasheet-pdf/pdf/219477/AKM/AK8973.html*) by Asahi Kasei Microsystems (available on Alldatasheet.com (*http://www.alldatasheet.com*)).

CHAPTER 4

HiJack

About This Chapter

Prerequisites
> Read Chapter 1 and the end of Chapter 2 (the section on the techBASIC help system) if you need some help with techBASIC. If you are already comfortable with tech-BASIC, jump right in.

Equipment
> You will need an iPhone, iPod, or iPad running iOS 5 or later. You will also need a HiJack device from Seeed Studio and various common electronics parts like wires, resistors, and a potentiometer. See Table 4-1 for a complete parts list.

> Newer iPhones and all iPads give out less power through the headphone port than the original devices HiJack was designed for. You may end up needing an external power supply. See Table 4-2 for the parts needed to build one kind of power supply.

Software
> You will need a copy of techBASIC or techBASIC Sampler.

What You Will Learn
> This chapter introduces HiJack, a device that supports an eight-bit analog-to-digital (A–D) converter plugged into the headphone port. It's the only wired technology Apple allows general-purpose apps to support.

What Is HiJack?

HiJack (shown in Figure 4-1) is a hardware device that plugs right into your iPhone, iPad, or iPod headphone jack. It's the only wired technology that can connect to iOS devices that isn't controlled by Apple's MFi program. This means it is also the only wired technology for which Apple will approve apps for the App Store without going through the MFi program.

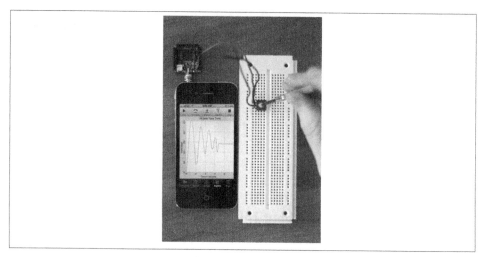

Figure 4-1. HiJack is a wired technology for sensor input

The MFi Program

Apple requires all devices that connect to the iPhone through the wired port at the bottom of the device or by classic Bluetooth to do so through the MFi program. This program requires the manufacturer to include a coded chip in each device. Apps that access the devices are universally rejected from the App Store unless they are preapproved by the MFi program, which almost always means the app was developed at the same time as the hardware and is the only app that works with the hardware. You can find out more about the MFi program at *https://developer.apple.com/programs/mfi/*.

This is actually why techBASIC cannot access devices like the Redpark Serial Cable. The developer and a few lucky beta testers have a version of techBASIC that works great with the Redpark Serial Cable, but Apple's MFi program would not give approval for a general-purpose technical programming language to support the cable.

This book explores other ways to access serial devices. Bluetooth low energy (Chapter 6) is not controlled by the MFi program like classic Bluetooth, and can be used to bridge to some devices, like the Android. WiFi bridges, like the WiFly device in Chapter 11, can also be used to access serial devices.

That's what makes HiJack so unique. It's not controlled by the MFi program, so programs like techBASIC are free to support wired connections with HiJack.

The current HiJack firmware supports an eight-bit A–D converter that takes an input of 0–2.75 volts. In this chapter we'll look at how to hook up the HiJack hardware and write a simple program to access data. The next chapter looks at a practical application.

In the programming world, it's traditional to start off with a new computer language by writing a simple program that prints "Hello world!" This convention started with Brian Kernighan, who wrote the first hello world program in a tutorial for the B programming language in 1972. With over 40 years of tradition behind us, we'll start our exploration of the HiJack hardware with a simple program to read the data, and a simple hardware project to provide that data—hello world, hardware style. We'll call our program "Hello HiJack."

One of the challenging things about interfacing hardware to a computer is that it involves at least two distinct disciplines: electrical engineering and programming. If you are reading this book, you're probably at least conversant with one of these fields, but very few people are expert in both. Because of that, I'll provide a lot more background in programming than the programmers need, and more background in electronics than the electrical engineers need. I'm sure you're good at skimming by now, so feel free to skip over the sections you already know.

Building the Sensor

Table 4-1 provides the list of parts you'll need for this project.

Table 4-1. Parts list

Part	Description
HiJack Development Pack	This includes everything you see in Figure 4-2.
Potentiometer	Anything in your parts bin will do. I used a 10KΩ potentiometer.
Breadboard	Breadboards are used to prototype electrical circuits. The white plastic part with all the holes that you saw back in Figure 4-1 is a popular size and style. We will use breadboards for many projects in this book. It's not required, but it makes life a lot easier.
Power supply	All iPads and all iPhones beginning with the iPhone 4S will require an external power source. This can be as simple as a 3-volt battery. A small, mobile power supply is shown in "External Power for HiJack" on page 61.

Let's start with the hardware itself. HiJack was developed at the University of Michigan for the purpose of creating cubic-inch sensor peripherals for mobile phones like the iPhone. It works just as well from the iPad. The whole idea is to give you a hardware platform that you can build on. You can buy the hardware from Seeed Studio (*http://www.seeedstudio.com/depot/*), which sells the raw hardware, a development pack, and a few odd components. I got the development pack, shown in Figure 4-2. It comes with the actual HiJack device, which is the green board with the headphone connection, along with a few other goodies.

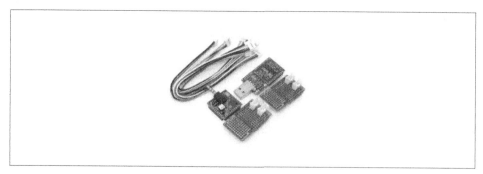

Figure 4-2. The HiJack development pack

The most important extra is the board with the USB plug; this is used to download new firmware as it becomes available. Be warned, though: the software to install firmware on the HiJack board only runs on Windows and is fairly temperamental. You won't have to do this right away—the HiJack device is preloaded with the current firmware—but if you ever want to update the firmware, a Windows computer is required.

The two blue boards are prototyping boards that are designed to plug snugly into the HiJack board. We'll make use of one of those later in the chapter. The various wiring harnesses are used to connect the prototyping boards to other devices, notably the prebuilt sensors in the Grove modular toolset. The Grove components include a collection of sensors, some of which work with HiJack right out of the box. They are also available from Seeed Studio. We'll be using one of them in the next chapter, so skip ahead and take a look at that before placing an order.

With a HiJack board in hand, the first thing we need to do is build a sensor. Ours will be about as simple as they come: we'll use a potentiometer (variable resistor) to provide a voltage we can change by varying the resistance. You can pick up potentiometers from pretty much anyplace that sells electronics components. The specific resistance doesn't matter, either. I used a tiny little 10KΩ (10,000 ohm) trim pot I had lying around in a parts bin.

The idea behind this sensor is to divide the voltage supplied by the HiJack device. Resistors in series divide an input voltage. Figure 4-3 shows a diagram of the circuit we'll be building. Purists will notice that I showed two separate resistors rather than a potentiometer. Electrically, they are the same thing as long as the potentiometer is not being adjusted, though, and it makes the following discussion a little easier.

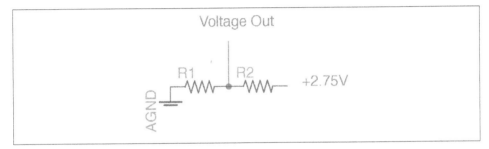

Figure 4-3. The HiJack Hello World circuit

The two lines labeled AGND at the left are the ground connection, abbreviated GND on the Seeed Studio documentation. This is the negative voltage side of the circuit. The other side, marked +2.75V, is the positive power supplied by the HiJack device. This pin is marked VCC on the Seeed Studio documentation. Our circuit will connect up to these two pins to draw power across the potentiometer.

A potentiometer is essentially, and sometimes literally, a resistive bar that uses a slider of some sort that covers the resistor. The bar is connected to a third wire. Sliding the bar changes the resistance to either side, dividing it into two parts that always add up to the total resistance in the device. The voltage output from the ground connection to the center wire of the resistor will be:

$$V_{out} = V_{in} \cdot \frac{R1}{R1 + R2}$$

Here, V_{in} is the 2.75 volts supplied by HiJack, and $R1$ and $R2$ vary as you adjust the potentiometer. Anything that detects the voltage from the ground connection to Voltage Out will see the voltage vary from 0 to 2.75 volts as the potentiometer is turned. That's exactly what the HiJack hardware will measure.

On the other end of the circuit sits the HiJack device itself. The top of the HiJack device has three female headers that provide 10 wiring connections each. Figure 4-4 shows the pin out, from a larger diagram available at the Seeed Studio site.

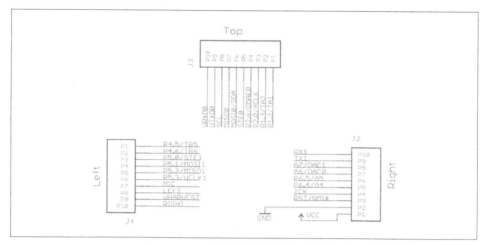

Figure 4-4. HiJack pin out

The connections we're interested in are on the right bar. We've already talked about GND and VCC; those pins provide the 2.75-volt power source we are using. The data input line is A6/DAC0. That's the pin we need to connect to the center pole on the potentiometer. Following the convention of black wires for ground and red for positive in DC circuits, and using a brown wire because I had one handy in my parts box, Figure 4-5 shows a closeup of the connections on the HiJack device.

Figure 4-5. Connections to HiJack

Potentiometers have three pins. In general, the center pin is the one we want to hook up to A6/DAC0, and GND and VCC can be connected to either of the two remaining pins. Figure 4-6 shows what it looks like in my circuit, with a second, identical potentiometer upside down beside the one I used so you can see the pins coming out of the bottom. I used a breadboard for the connections, but that's because I had one lying around. Anything that will securely mount the wires to the potentiometer and let you easily adjust the resistance will work just fine.

Figure 4-6. Connections to the potentiometer

External Power for HiJack

Table 4-2 provides the parts list for the external power supply.

Table 4-2. Parts for external power supply

Part	Description
Prototype board	This is one of the two prototype boards that come with the HiJack Development Pack.
Battery holder	Jameco part 17002-00603 or similar.
2032 battery	These are available from lots of places, but you might want to buy 10 or so when ordering other parts. They'll come in handy for the Bluetooth low energy projects later in the book.
Switch	Any switch that has pin spacing in multiples of 2 mm will do. I was lucky enough to find an old slide switch with 4 mm pin spacing in my parts box, but I can't find any replacements. You may need to bend the pins a bit on a switch designed for a different spacing.
Resistors	One should be 10 times the resistance of the other. I used a 220Ω and a 2.2KΩ.

HiJack was designed to draw power from the audio output of the headphone port, a clever design feature that started to fail when Apple reduced the power output beginning with the iPhone 4S and on all models of the iPad. It's easy enough to correct this problem by supplying 2.75V to VCC on the HiJack chip, as shown in Figure 4-7. The designers strongly cautioned me, however, not to exceed 3V.

Figure 4-7. The circuit diagram and circuit for a simple HiJack power supply

 Warning: Don't exceed 3V!
Providing more than 3V from an external power source can damage the HiJack device. Since most 3V batteries will actually supply 3.3V when new, I was overly cautious and dropped the voltage using a resistive circuit very much like the one in the Hello World circuit. If you supply external power to a HiJack, please limit the voltage using a simple circuit like this one or, better still, a regulated power supply.

One way to supply the power is with a coin cell battery attached to one of the prototyping boards that comes with the HiJack development kit, as shown in Figure 4-8.

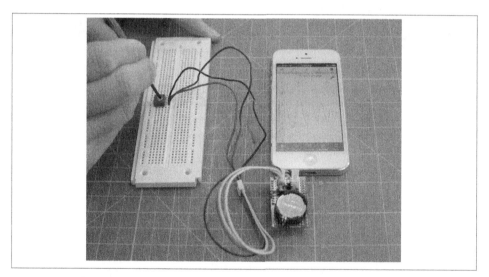

Figure 4-8. The Hello World circuit using the external power supply

You can get really fancy or keep it simple. Due to the limited working space, I chose simplicity, using a 220Ω resistor and a 2,200Ω resistor to drop the voltage from 3.3V to 3V and a switch to turn the power on or off. Mounting the circuit on one of the prototyping boards makes for a very convenient, portable package that works very well with the Grove sensors. It's easy to tap A6, VCC, and GND from the connector on top of the prototyping board using one of the cables that comes with the HiJack Development Pack. The black and red wires are ground and +2.75V, respectively, and the yellow wire is A6.

Hello HiJack

With all of that hooked up, it's time to write our program (shown running in Figure 4-9) to display what's happening.

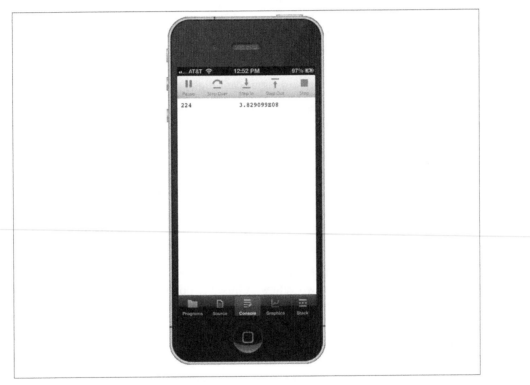

Figure 4-9. The Hello HiJack program

Run techBASIC from your iPhone or iPad, tap the New button to create a new program, and enter "Hello HiJack" as the program name. Tap the Source screen to get the keyboard and enter this program:

```
WHILE 1 ❶
    System.clearConsole ❷
    PRINT HiJack.receive ❸
    System.wait(0.5) ❹
WEND
```

Here's how the program works:

❶ The `WHILE` and `WEND` lines form a loop that will continue as long as the expression in the `WHILE` statement is nonzero. One will be nonzero for a really long time, so the loop will go forever. The only way to stop this program will be to press the Stop button that will show up at the top of the screen when the program starts running.

❷ `System.clearConsole` clears any text from the console, where text input and output appear.

❸ The `PRINT` statement then prints a two-dimensional array to the screen. It gets that array from the HiJack hardware--`HiJack.receive` is a function that returns an array where the first element is a value between 0 and 255, with lower numbers when the voltage is low and higher numbers when it is high. The second value on the line is a timestamp. The timestamp may not seem to change much, but it's a large number whose most significant digits change slowly.

❹ `System.wait(0.5)` pauses for a moment so we get a chance to read the screen without too much flicker. The parameter tells the call how long to pause, in this case for half a second.

After typing in the program, plug the HiJack hardware into the headphone jack on your iPhone. Tap the Programs button at the bottom of the screen, then tap the name of the program. techBASIC will change the display to the console, where you should see a number that occasionally changes. The output will look something like this:

```
71              3.804955E08
```

After a second or two, the first number will settle down to a single value, or perhaps flick back and forth between two values. Adjust the potentiometer, and you'll see the number change. Just a few more steps, and you'll be building RoboCop!

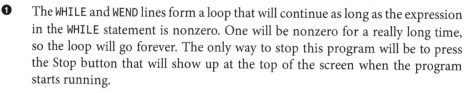

How Long Will 1 Be 1?

There is a quip in the text about 1 being nonzero for a really long time. "Ha, ha," you say. "That's always true." And yet...

Back when I first started programming, we were working on an IBM 360 mainframe using FORTRAN IV. After a really long time debugging a program, one of the programmers realized the only way the program could behave as it did was if 1 was, well, not 1. After still more searching, the reason was discovered. To make programs faster, the FORTRAN compiler took advantage of the fact that it took less time to load a value from memory than to load a constant, so it stored the value 1 in memory and loaded it when needed. FORTRAN IV also happened to pass all variables by reference, which meant changing the value in a subroutine changed the original value, not a local copy.

Someone had passed 1 to a subroutine instead of I, and the subroutine had changed the value of 1!

The moral is never to assume anything when debugging. Oh, sure, start with the obvious stuff, but in the end, check every possibility—even the ones you *know* can't be the problem.

When Things Go Wrong

OK, not so fast. Did it work? If not, there are two places for things to go wrong: the software or the hardware.

If there is a problem with the software, techBASIC may say so. Check to make sure the program is exactly the one typed above. Letter case is not important, but spaces and line breaks can be. Make sure there are five lines, and that there are spaces between the words where shown, and no extra spaces inserted. `clear Console` is not the same as `clearConsole`! Also, look at the error messages from techBASIC. They will usually pinpoint the problem, and even if they don't, the real problem will be close by.

If the program is running and printing numbers, but the numbers seem to jump around randomly, you probably don't have a good connection or the HiJack device does not have enough power. Check to make sure the HiJack hardware is plugged all the way into the headphone port. The metal disk on the washer does not fit flush against the case, so don't push too hard. If you are using an iPod Touch, you may also need to twist the jack around a bit.

Power can be a real problem with the HiJack device. There is no battery—it is getting its power from an audio tone generated by the iPhone. Make sure the volume is turned all the way up. Also, Apple lowered the power output for the headphones starting with the iPhone 4S and with all models of the iPad. You'll need an external power source for the later model devices (refer back to "External Power for HiJack" on page 61). With an external power supply, HiJack doesn't need to draw power from the sound coming out of the headphone port, so you can turn it down. Leaving the volume turned up won't do any harm, but it will drain the iPhone's battery a little faster.

The last potential source of trouble is the HiJack hardware or the circuit. Check it carefully to make sure the right wires are connected to the right places, and that the wires are making a good electrical contact.

A Better HiJack Program

While that first little program worked, it's not the most exciting program in the world. Wouldn't it be nicer to have something like an oscilloscope trace, like the ones from the programs in the first three chapters that access the internal sensors? It's actually not that hard. In fact, let's do that now.

Our goal is a program that plots results like those in Figure 4-10. We'll plot 10 seconds' worth of data, collecting a data point every 0.1 seconds, for a total of 100 data points plotted at any one time. The newest data will always appear at the right, at time=0, and older data will be scrolled to the left, with time getting more and more negative until the point falls off of the display.

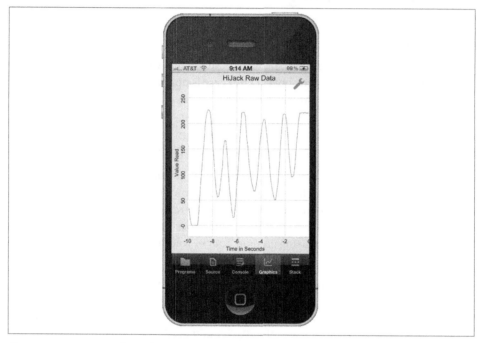

Figure 4-10. Output from the GUI-based HiJack program

Let's take a look at the code. I'll show the complete program first, then we'll walk through it line by line.

 This program is included in techBASIC and techBASIC Sampler. Look for the app called HiJack in the *O'Reilly Books* folder.

The finished program looks like this:

```
! Shows a running plot of HiJack
! input for the last 10 seconds
! in 0.1-second intervals.
```

```
!
! Initialize the display with the
! value set to 0.
DIM value(100, 2)
FOR t = 1 TO 100
  value(t, 1) = (t - 100)/10.0
NEXT

! Initialize the plot and show
! it.
DIM p as Plot, ph as PlotPoint
p = Graphics.newPlot
p.setTitle("HiJack Raw Data")
p.setXAxisLabel("Time in Seconds")
p.setYAxisLabel("Value Read")
p.showGrid(1)
p.setGridColor(0.8, 0.8, 0.8)

ph = p.newPlot(value)
ph.setColor(1, 0, 0)
ph.setPointColor(1, 0, 0)

! Set the plot range and
! domain. This must be done
! after adding the first
! PlotPoint, since that also
! sets the range and domain.
p.setView(-10, 0, 0, 255, 0)

system.showGraphics

! Loop continuously, collecting
! HiJack data and updating the
! plot.
DIM time AS double
time = System.ticks - 10.0
WHILE 1
  ! Wait for 0.1 seconds to
  ! elapse.
  WHILE System.ticks < time + 10.1
  WEND
  time = time + 0.1

  ! Get and plot one data point.
  h = HiJack.receive
  FOR i = 1 TO 99
    value(i, 2) = value(i + 1, 2)
  NEXT
  value(100, 2) = h(1)
  ph.setPoints(value)
  Graphics.repaint
WEND
```

OK, that's not too long, as programs go. It's just 55 lines, and a lot of them are comments. Let's see what it does. Here's the first chunk:

```
! Shows a running plot of HiJack
! input for the last 10 seconds
! in 0.1-second intervals.
!
! Initialize the display with the
! value set to 0.
DIM value(100, 2)
FOR t = 1 TO 100
  value(t, 1) = (t - 100)/10.0
NEXT
```

This first chunk of code makes some introductory comments, then sets up an array to hold the values we will eventually read from the HiJack hardware. The array value will hold up to 100 values. It's a two-dimensional array because we will need to tell techBASIC both the x and y values for each point to plot. The x values are the timeline, which doesn't change, so we use a FOR loop to fill in these values. Our intent is to collect one point every 0.1 seconds and display 10 seconds' worth of data, so we fill in the x values with values ranging from -9.9 to 0. We can safely leave the y values unchanged, since BASIC initializes new variables to 0, and that will work fine for our purposes. All of this will sound very familiar if you have already worked through the first three chapters.

```
! Initialize the plot and show
! it.
DIM p as Plot, ph as PlotPoint
```

Next, we need to create a plot. The DIM statement sets up two variables, one to hold the Plot class that displays the plot itself and another for the PlotPoint class, which contains the actual points to plot.

```
p = Graphics.newPlot
p.setTitle("HiJack Raw Data")
p.setXAxisLabel("Time in Seconds")
p.setYAxisLabel("Value Read")
p.showGrid(1)
p.setGridColor(0.8, 0.8, 0.8)
```

The first line of the next block creates the plot itself. Think of this as the background, including the titles, axes, and so forth. The setTitle method sets the title at the top of the plot, while the next two lines set the axis labels. showGrid turns on the grid lines that appear behind the plot line; without this call, the background is blank. Finally, we set the grid color to a light gray. Like almost all techBASIC calls that take a color, setGrid Color takes three parameters, one each for the intensity of the red, green, and blue colors, in that order. The valid values range from 0.0 to 1.0, with 0.0 being black and 1.0 being the full intensity for that color. A few calls have a fourth number for the alpha channel, which tells how transparent a color is. That lets you draw something over a background and, to the degree specified by the alpha value, see through the new color to whatever lies behind it. Check out the RGB color model article on Wikipedia (*http://en.wikipedia.org/wiki/RGB_color_model*) if you would like to know more about how RGB color works on a computer.

```
ph = p.newPlot(value)
ph.setColor(1, 0, 0)
ph.setPointColor(1, 0, 0)
```

The next step is to set up the PlotPoint class that actually draws the line across the plot. The newPlot method creates a new instance of the class and adds it to the plot we just created. We then set the color for both the line and the points where the data is actually plotted to red. There are other calls in the PlotPoint class that control the shape of the points and line; you can use those to customize your version of the program.

```
! Set the plot range and
! domain. This must be done
! after adding the first
! PlotPoint, since that also
! sets the range and domain.
p.setView(-10, 0, 0, 255, 0)
```

HiJack always returns a value from 0 to 255, and we know the x-axis will show times from –10 to 0 seconds, so next we set the axis to show exactly those values. Without this call, techBASIC will default to showing 0 to 10 along the x-axis and roughly –5 to 5 along the y-axis. We can always change what we are looking at with some swipe and pinch gestures, but this saves us the trouble.

```
system.showGraphics
```

Now that things are set up, we tell techBASIC to switch to the graphics display. Again, we could have done this with a tap on the Graphics button, but this saves us the trouble.

```
! Loop continuously, collecting
! HiJack data and updating the
! plot.
DIM time AS double
time = System.ticks - 10.0
```

HiJack can report data at various rates, up to a little more than 100 points per second. The default rate in techBASIC is about 40 values per second, which is more than we need. These lines set up a timestamp we will use to tell when 0.1 seconds have elapsed. Each time that happens, we'll grab a new point from the HiJack hardware and add it to our plot.

```
WHILE 1
```

Just like the first program, this program will loop until you manually end it with the Stop button.

```
! Wait for 0.1 seconds to
! elapse.
```

```
WHILE System.ticks < time + 10.1
WEND
time = time + 0.1
```

Here is where we wait for 0.1 seconds to elapse. The WHILE loop waits until the system clock reports a time 10.1 seconds past the original time we recorded before the loop started. We then add 0.1 to this time so the next time through, this timer loop will wait until 10.2 seconds have gone by, and so forth.

```
! Get and plot one data point.
h = HiJack.receive
```

This is all it takes to actually read the HiJack hardware. A value from 0 to 255 is stuffed into the variable h.

```
FOR i = 1 TO 99
  value(i, 2) = value(i + 1, 2)
NEXT
```

This loop shifts the 99 most recent points in the value array one index lower, which will cause them to be drawn one point to the left on the plot. Remember, the time is preset and is not being shifted, so this essentially makes each point 0.1 seconds older on the plot. The first time through, this is just copying a bunch of zeros, but after 10 seconds, all of the values are older values read from the HiJack hardware.

```
value(100, 2) = h(1)
```

The new point goes in the last spot in the plot. We pull off the first element of the array, which is the HiJack data, and ignore the second, which contains a timestamp we don't need in this program.

```
ph.setPoints(value)
Graphics.repaint
```

This tells the PlotPoint object we created and stored in ph to use a new set of points. Next we repaint the graphics screen, showing the new information on the plot.

```
WEND
```

Finally, we go back to the WHILE statement and do it all again.

Try out the finished program with HiJack, as shown in Figure 4-11. You should be able to see exactly where the potentiometer is set, and watch the change as you adjust it. This program will work with all of your HiJack projects, although you will probably develop custom programs for specific sensors and uses.

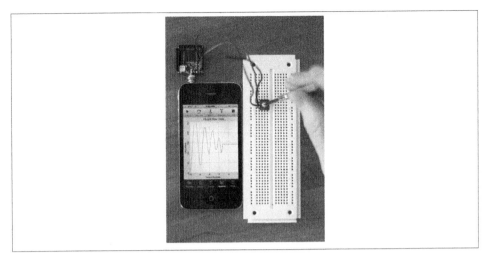

Figure 4-11. The completed HiJack project

For More Information

You can find out more about HiJack at the University of Michigan's HiJack page (*http://web.eecs.umich.edu/~prabal/projects/hijack/*) and at the wiki on the Seeed Studio site (*http://www.seeedstudio.com/wiki/index.php?title=Hijack*). The Seeed Studio website also has information about other hardware projects you can build, downloads for the software to update the firmware on the HiJack, and the latest version of the firmware.

CHAPTER 5

Creating a Moisture Meter with HiJack

About This Chapter

Prerequisites

Read Chapter 1 and the end of Chapter 2 (the section on the techBASIC help system) if you need some help with techBASIC. Read Chapter 4 for general information about HiJack. In particular, iPads and late model iPhones need an external power supply for HiJack; see Chapter 4 for details.

Equipment

You will need an iPhone, iPod, or iPad running iOS 5 or later. You will also need a HiJack device from Seeed Studio, a Grove moisture meter (also available from Seeed Studio), and various common electronics parts like wires and resistors. See Table 5-1 for a complete parts list.

Software

You will need a copy of techBASIC or techBASIC Sampler.

What You Will Learn

This chapter uses HiJack, a device that supports an eight-bit A-D converter plugged into the headphone port, to build a plant moisture meter. This is a functioning example that shows how to use the technology introduced in Chapter 4 in a real-world sensor application.

Adding a Moisture Meter to the Tricorder

Back in Chapter 1 and Chapter 2, we explored the tricorder we all carry in our pocket. This project turns an iPhone or iPad into a *quadcorder*, hanging a HiJack A–D converter and sensor off the end and then using techBASIC to read it. We're going to build a nifty moisture meter from off-the-shelf parts, as shown in Figure 5-1. Even the initial software is something you've seen before: we'll start with the HiJack program from Figure 4-10 in Chapter 4.

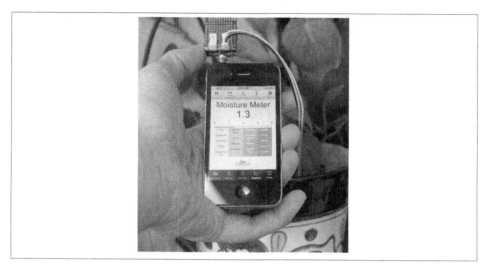

Figure 5-1. Moisture meter

Because all of the components, even the software, are off-the-shelf, we can concentrate on the part of the project we're best at—or, for those of us perverse enough to think this way, the part we're worst at so we can improve. The whole project can be tossed together in a matter of minutes. Later we'll develop a much cooler implementation of the software so our quadcorder's moisture meter is easier to use, and looks a whole lot better, too.

Assembling the Moisture Meter

Table 5-1 provides a list of parts needed for the moisture meter.

Table 5-1. Parts list

Part	Description
iPhone or iPad	Any iPhone or iPad that will run iOS 5.
HiJack	The HiJack A-D converter, available from Seeed Studio (*http://www.seeedstudio.com*).
External power	If you are using an iPhone 4 or later or an iPad, you will need to add an external power supply. See External Power for HiJack in Chapter 4 for one solution.
Moisture sensor	Seeed Studio sells a series of components under the name Grove. These are actually part of another development system, but they share the same harness and electrical characteristics as the HiJack. The last piece of hardware we will need for this project is the Grove moisture sensor, available for $4.99. Be sure to check out the other Grove components while you are on the Seeed Studio site; you may be surprised at the number that are available.

Figure 5-2 shows exploded and assembled views of the hardware. All of the parts except the moisture sensor come with the HiJack Development Pack. It really is as simple as it looks—just plug the components together as shown, then insert the headphone jack into an iPhone.

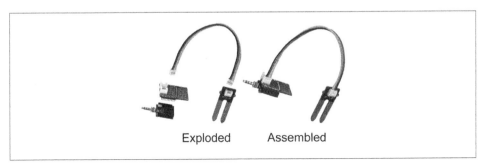

Exploded   Assembled

Figure 5-2. Exploded and assembled views of the hardware

The HiJack software from Figure 4-10 in Chapter 4 works with any HiJack component. Plug the assembled moisture sensor into your iPhone or iPad and run the HiJack software by tapping the program name. You should get different readings from dry air, pressing your fingers on the moisture sensor, or pushing it into potted plant soil with various amounts of moisture in it, as shown in Figure 5-3.

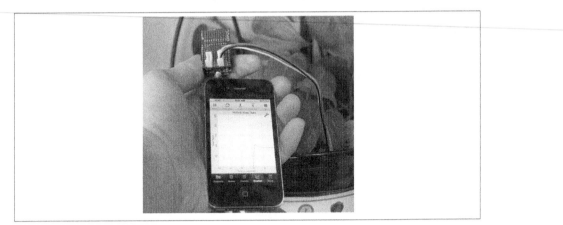

Figure 5-3. HiJack software and hardware with the Grove moisture sensor

Calibration

That's fine, as far as it goes, but what do the values mean? Calibration is an important part of creating any measurement instrument, and this one is no different. To calibrate the moisture meter, I compared it to a commercial plant moisture meter, which lists the moisture in a simplified range of one to four. You can skip this messy but fun step and accept my calibration values, or have fun playing in the mud to do your own calibration.

Collecting the Calibration Data

As shown in Figure 5-4, I put 150 grams of potting soil into a one-liter container and stirred in small amounts of water in 10 cc increments, measuring the moisture in the soil with both the commercial moisture meter and the HiJack moisture meter. Be sure to pack the soil before taking a measurement. If you leave the soil loose after stirring in the water, the soil doesn't make good contact with the moisture sensors, and the readings are very low and inconsistent.

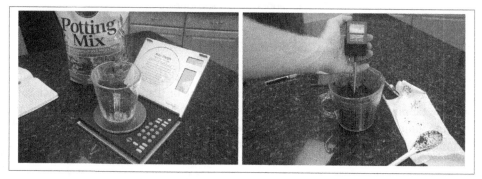

Figure 5-4. Calibrating the moisture meter

I used a spreadsheet to record the data, and then saved it as a CSV (comma-separated values) file. My data looks like this:

```
81,0.1
94,0.5
100,1.5
103,2
109,2
112,2.2
112,2.3
79,1
103,2.5
72,1
105,2.5
110,3.8
111,3.8
80,1.5
86,1.5
97,3
105,3
110,3.5
```

```
111,3.5
113,4
113,4
118,4
118,4
118,4.5
120,4.5
77,0.8
76,0.8
92,1.2
97,1.2
92,2
105,2
110,2.2
108,2.2
113,3.2
115,3.2
113,3.5
114,3.5
114,4
116,4
```

The file itself is already loaded in techBASIC and techBASIC Sampler; it is called *moisture.csv*.

 CSV is a very common file format for moving and storing information as text files, and it works great for tables of numbers. You could use a text editor instead of a spreadsheet, but most spreadsheets can export CSV files.

Moving Datafiles to and from techBASIC

The *moisture.csv* file comes with techBASIC, but what if you want to use your own data? It turns out it's not really hard to move datafiles back and forth between your desktop computer and techBASIC, as shown in Figure 5-5.

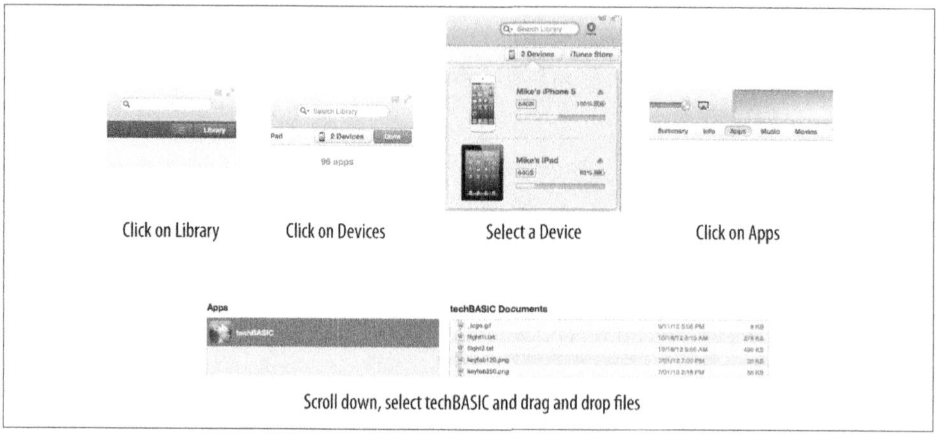

Click on Library   Click on Devices   Select a Device   Click on Apps

Scroll down, select techBASIC and drag and drop files

Figure 5-5. Moving datafiles to and from techBASIC

Here are the steps to move files to and from techBASIC using iTunes:

- From iTunes, click the Library button at the top right of the page. This tells iTunes to show the stuff you have rather than the stuff you can buy on the iTunes Store.

- Click on the Devices button to get a list of devices. Select the device you want to move datafiles to or from.

- Click on Apps in the menu bar near the top of the window. This shows the apps on the device.

- Scroll down until you see the Apps list, and select techBASIC. The panel just to the right of the Apps list changes to techBASIC Documents. This is a list of all of the datafiles on the device.

- Drag and drop files to and from the techBASIC Documents panel to add or retrieve datafiles. Select a file and press Delete to remove it from the device.

It's fair to ask why you can't move techBASIC programs to and from your iPad or iPhone the same way. At one time, you could—but Apple required the removal of this feature from techBASIC. While you can move a program file to the techBASIC datafile area, techBASIC will not recognize the file as a BASIC program. Check back from time to time, though. If this restriction is ever removed, techBASIC will allow moving program files again.

Using the Calibration Data

Looking at the data shown in Figure 5-6, it's clear that the relationship is fairly linear—certainly close enough to linear for the purpose of calibrating a moisture meter that will convert 256 moisture sensor readings to 4 moisture values.

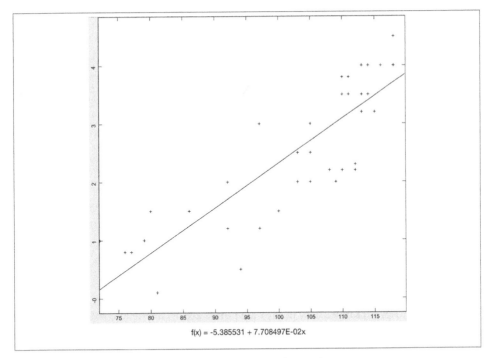

$f(x) = -5.385531 + 7.708497E\text{-}02x$

Figure 5-6. HiJack (x-axis) versus moisture meter (y-axis)

The best and easiest way to use this information to convert from HiJack data to moisture meter equivalents is by using linear regression, a technique for finding the line that gives the lowest error for all of the measurements. Fortunately, linear regression is built right into techBASIC—that's an advantage of using a technical computing environment for our programs. Here is the program I used. It's also in techBASIC and techBASIC Sampler in the *O'Reilly Books* folder—look for the program named Moisture Calibration.

```
! Perform linear regression on a CSV file. Each
! line of the file should contain an X and Y
! value separated by a comma.
!
! Determine the number of values. Also find the
! min and max values for X, used later.
name$ = "moisture.csv"
OPEN name$ FOR INPUT AS #1
n = 0
minX = 1e50
maxX = -minX
WHILE NOT EOF(1)
   INPUT #1, x0, y0
   IF x0 < minX THEN minX = x0
   IF x0 > maxX THEN maxX = x0
```

```
    n = n + 1
WEND
CLOSE #1
```

The first step is to scan the datafile and count the number of data points. Each data point is on a separate line, so the program reads the lines from the file and counts them. We'll need the minimum and maximum values for *x* later, when we create a plot of the data showing the fit, so that is collected here, too.

```
! Dimension arrays for the values. The x
! and y arrays are used for regression,
! while the v array is used for the plot.
DIM v(n, 2), x(n), y(n)

! Read the values.
OPEN name$ FOR INPUT AS #1
FOR i = 1 TO n
    INPUT #1, x(i), y(i)
    v(i, 1) = x(i)
    v(i, 2) = y(i)
NEXT
CLOSE #1
```

Now that the program knows how many data points are in the file, it creates three arrays to hold them and reads the data. Why three arrays? The `Math.poly` method in techBA-SIC that performs the actual regression needs one array of *x* values and another of *y* values. The `PlotPoint.newPlot` method that creates a scatter plot of the data uses a single array of *x-y* data pairs. The easiest way to satisfy both requirements is to build the arrays separately.

```
! Do the regression.
coef = Math.polyfit(x, y)
```

Doing the regression is the easy part. `Math.poly` returns an array with two elements. The first is the *y* intercept, while the second is the slope. For any *x* value we choose, the equivalent *y* value is found like this:

```
y = coef(0) + coef(1)*x
```

The actual values are −5.385531 and 0.07708497. This allowed me to convert my HiJack readings into the equivalent moisture meter readings using this equation:

$$m = 5.385531 + 0.07708497h$$

where *h* is the HiJack reading and *m* is the equivalent reading from the commercial moisture meter.

```
! Create an array showing the fit.
DIM fit(0 TO 10, 2)
FOR i = 0 TO 10
    fit(i, 1) = minX + i*(maxX - minX)/10
    fit(i, 2) = coef(1) + coef(2)*fit(i, 1)
```

```
NEXT

! Create the plot. Add the individual points
! and the fitted line.
DIM myPlot AS Plot, scatterPlot AS PlotPoint, fitPlot AS PlotPoint
myPlot = Graphics.newPlot
scatterPlot = myPlot.newPlot(v)
scatterPlot.setStyle(0)
scatterPlot.setPointStyle(2)
fitPlot = myPlot.newPlot(fit)
myPlot.setRect(0, 0, Graphics.width, Graphics.height - 41)

! Add a label showing the equation of the fit.
DIM equation AS Label
equation = Graphics.newLabel(0, Graphics.height - 31, Graphics.width)
equation.setAlignment(2)
e$ = "f(x) = " & STR(coef(1)) & " + " & STR(coef(2)) & "x"
equation.setText(e$)

! Show the graphics screen.
System.showGraphics
```

The rest of the program creates and displays the plot. It will be a bit shorter when it first displays; use a pinch gesture to expand the y-axis to duplicate the figure.

According to the directions on the commercial moisture meter, it's time to water a plant when the soil hits the watering number for the plant. For example, orchids need fairly dry soil, so they are not watered until the moisture meter reports a reading of less than 1.

Better Software

While the general HiJack software works, it doesn't have much pizzazz. We'll create a custom program (shown in Figure 5-7) to read the moisture meter data and present it in a much more pleasing way. As is typical with this sort of program, most of the code will be for setting up the graphical user interface. The actual code to read HiJack is just one line in the program.

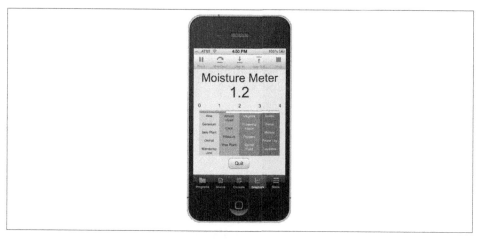

Figure 5-7. The moisture meter GUI

Let's look at the design for a moment before looking at how the program is written. The design serves as a roadmap for discussing the code.

A large label reminding us what this program does occupies the top of the screen. Below that is the digital moisture value in another label. We'll use a progress bar as an analog moisture meter, seen here right below five labels used to show the scale.

There are four TextView objects below the progress bar. These are all art, but useful art. Color is used to show the relative moisture, starting with light blue for dry soil and moving to darker blue for wet soil. A few common plants are listed in each group; when the soil's moisture is at or below the level for a particular plant, it's time to water the plant. Finally, there is a Quit button at the bottom of the screen to stop the program.

Now we'll step through the program to see how it works. For discussion purposes, the subroutines and functions appear as they are introduced, intermixed with the main body of the program, but in a completed program they should appear at the end. You'll find the complete program at the end of the chapter, and in the *O'Reilly Books* folder in both techBASIC and techBASIC Sampler.

```
! HiJack Moisture Meter

! Show the graphics screen
System.showGraphics(1)
System.setAllowedOrientations(1)
Graphics.setToolsHidden(1)

! Get the size of the graphics
! screen
width = Graphics.width
height = Graphics.height
```

We'll use the size of the graphics screen to calculate appropriate values for the position and size of controls. To save some typing, the program starts by placing these values in local variables. We're also going to use the full screen and hide the tools button that provides access to the debugger. Since showing the full screen changes the size of the graphics area, the program does this right away.

Of course, if there is a bug in the program, this leaves us with no way to stop it. During debugging or development, change these lines to:

```
! HiJack Moisture Meter

! Show the graphics screen
System.showGraphics(0)
System.setAllowedOrientations(1)
!Graphics.setToolsHidden(1)

! Get the size of the graphics
! screen
width = Graphics.width
height = Graphics.height
```

This will keep the tools button and display the techBASIC controls along the top and bottom, making it possible to debug or interrupt the program.

```
! Paint the background light gray
bg = 0.9
Graphics.setColor(bg, bg, bg)
Graphics.fillRect(0, 0, width, height)
```

The default screen is white, and that just won't do. techBASIC has a built-in class called Graphics that is used to draw on the graphics screen. setColor and fillRect paint the entire screen a light gray. You can adjust the shade of gray by changing the value of bg.

```
! Create a Quit button
DIM quit AS Button
quit = Graphics.newButton(width/2 - 36, height - 57)
quit.setTitle("Quit")
quit.setBackgroundColor(1, 1, 1)
quit.setGradientColor(0.7, 0.7, 0.7)
```

Our program will be an event-driven program, so it will run until it is stopped. Since our program uses the full screen, the techBASIC button normally used to stop a program won't be available. These lines create a stop button centered near the bottom of the screen. Rather than the default white button, we're using a gradient to create a shadowed button.

To make the stop button function, we'll need to add a subroutine that handles button clicks. Here's the one in our program, found near the bottom of the complete listing. It checks to make sure it was the Quit button that was tapped, then stops the program:

```
! Handle a tap on a button
!
! Parameters:
!   ctrl - The button tapped
!   time - When the button was
!        tapped

SUB touchUpInside(ctrl AS Button, time AS DOUBLE)
IF ctrl = quit THEN
  STOP
END IF
END SUB
```

This is enough code to produce a working program. When you run it, you should see a Quit button on a gray background, and tapping the Quit button should exit the program.

```
! Put the name of the program at
! the top of the screen
DIM mmLabel AS Label
mmLabel = newLabel(0, 20, width, 40, 40, "Moisture Meter")
mmLabel.setBackgroundColor(bg, bg, bg)
```

This code creates a label. Labels are small chunks of text, usually used to label other components. We'll be creating a lot of labels with various positions, sizes, and text, so this code is actually calling a subroutine in our program to do some of the repetitive work, followed by setting the background color for the label so it matches our screen background. Here's the subroutine that is called to create the label; it appears at the end of the complete program with the rest of the subroutines and functions:

```
! Create a label
!
! Parameters:
!   x - Horizontal location
!   y - Vertical location
!   width - Label width
!   height - Label height
!   fontSize - Point size for the
!        font
!   text$ - Label text
!
! Returns: The label

FUNCTION newLabel (x, y, width, height, fontSize, text$) AS Label
DIM nl AS Label
nl = Graphics.newLabel(x, y, width, height)
nl.setText(text$)
nl.setBackgroundColor(1, 1, 1, 0)
nl.setAlignment(2)
nl.setFont("Sans_Serif", fontSize, 0)
newLabel = nl
END FUNCTION
```

This subroutine creates a variable called nl, short for new label, to hold the label, then calls Graphics.newLabel to create the actual label. nl.setText sets the text for the label.

nl.setBackgroundColor sets the background color using the normal three red, green, and blue components, whose values range from 0 to 1, but in this case, it also sets the alpha level. The alpha level controls how opaque the color is. By setting the background alpha level to 0, we're setting it to be completely transparent so anything under the label shows through. This means we don't have to set the background color for each label to the background screen color. But isn't that exactly what we did after creating the mmLabel a moment ago? Well, yes—but that was a special case. It turns out techBASIC puts a control on the graphics screen to give you some options for dealing with plots. Since we can't use the button if the label is on top of it, we make the label opaque to hide the button. An alternative would be to use the techBASIC Graphics.setToolsHidden method to hide the tools. Either way works.

The next two lines center the text and set the font size, using the system's default sans-serif font. Finally, we set the return value and return the new label to the caller.

```
! Create a large label to show
! the moisture level
DIM value AS Label
value = newLabel(0, 75, width, 40, 50, "0")
```

Back in the main program, the same newLabel subroutine is now used to create a large label that will display the digital readout for the moisture meter. We'll see the subroutine that actually sets the value later. For now, we start with a reading of 0.

```
! Add 5 small labels to show the
! moisture scale along the top of
! the moisture bar
DIM nums(5) AS Label
plantLabelWidth = (width - 40)/4
FOR i = 0 TO 4
  x = i*plantLabelWidth
  nums(i + 1) = newLabel(x, 140, 40, 20, 16, STR(i))
NEXT
```

Our newLabel subroutine is getting quite a workout! Here we use it again to create five labels, 0 to 4, which show the scale for the analog readout. The various calculations evenly space the five labels across an area of the screen that extends from 20 pixels from the left edge to 20 pixels from the right edge, which is the size we'll use in a moment for the progress bar we'll use as an analog meter.

```
! Create the strings that will
! name the plants in each
! moisture group
DIM plants(4) AS TextView, plants$(4)
addPlant("Aloe", plants$(1))
addPlant("Geranium", plants$(1))
```

```
addPlant("Jade Plant", plants$(1))
addPlant("Orchid", plants$(1))
addPlant("Wandering Jew", plants$(1))
addPlant("African Violet", plants$(2))
addPlant("Cacti", plants$(2))
addPlant("Hibiscus", plants$(2))
addPlant("Wax Plant", plants$(2))
addPlant("Begonia", plants$(3))
addPlant("Flowering Maple", plants$(3))
addPlant("Peppers", plants$(3))
addPlant("Spider Plant", plants$(3))
addPlant("Azalea", plants$(4))
addPlant("Ferns", plants$(4))
addPlant("Melons", plants$(4))
addPlant("Peace Lily", plants$(4))
addPlant("Tomatoes", plants$(4))
```

We're going to add four text views now, each of which will have a background color that indicates the relative moisture level, and each of which will have a list of common plants that should be watered when the soil is at or below the indicated moisture level. These will slightly overlap the progress bar, so we want to create them first so the progress bar is drawn on top. This makes them look like an integral part of the analog meter, rather than an afterthought sitting below it.

This first chunk of code sets up the text that will appear in each text view. It calls the addPlant subroutine that appears at the end of the program with the other subroutines and functions.

```
! Add a plant name to a string
! containing plant names
!
! Parameters:
!   newPlant$ - New plant name
!   plant$ - Current plant names

SUB addPlant (newPlant$, BYREF plant$)
IF LEN(plant$) <> 0 THEN
  plant$ = plant$ & CHR(10) & CHR(10)
END IF
plant$ = plant$ & newPlant$
END SUB
```

This subroutine checks to see if the list of plants is empty. If not, it adds two new line characters to the string, then adds the new plant name. Parameters can be passed by value or by reference. Passing a parameter by value creates a local copy in the subroutine, so changing the value of the parameter does not change the value passed to the subroutine. Passing a parameter by reference passes the original variable, so changing the parameter in the subroutine does change the original value. Since the plant$ parameter is passed by reference, the original value in the program is updated by the call without the need to create a function.

```
! Add colored labels below the
! moisture bar showing the plants
! in each group
plantLabelHeight = 150
FOR i = 1 TO 4
  x = 20 + (i - 1)*plantLabelWidth
  color = 1 - i/5
  plants(i) = newTextView(x, 170, plantLabelWidth, plantLabelHeight, _
                      11, color, plants$(i))
NEXT
```

Next we create the four text views. There's a bit of algebra required to make them fit evenly across the screen and to set the color, but most of the work is done in the newTextView subroutine. The color we're setting is actually the white level for the background of the text view, so it's brighter for the low-moisture text views. We'll see how this is used as we work through the newTextView subroutine, again collected here from later in the complete program listing:

```
! Create a text view to show a
! list of plants
!
! Parameters:
!   x - Horizontal location
!   y - Vertical location
!   width - TextView width
!   height - TextView height
!   fontSize - Point size for the
!       font
!   color - White level for
!       background; the color will
!       be blue, lightened by this
!       amount
!   text$ - TextView text
!
! Returns: The text view

FUNCTION newTextView (x, y, width, height, fontSize, color, text$) AS TextView
DIM ntv AS TextView
ntv = Graphics.newTextView(x, y, width, height)
ntv.setText(text$)
ntv.setEditable(0)
ntv.setBackgroundColor(color, color, 1, 1)
IF color < 0.5 THEN
  ntv.setColor(1, 1, 1)
END IF
ntv.setAlignment(2)
ntv.setFont("Sans_Serif", fontSize, 0)
newTextView = ntv
END FUNCTION
```

Most of the newTextView subroutine should look familiar, since it's very similar to the newLabel subroutine we looked at earlier. Other than returning a text view instead of a label, there are really only two differences. The first is ntv.setEditable, which tells the control that the user can't edit the text. The other difference is the way the color is set. In this case, we set the red and green components of the background color to the value passed as the color parameter, and set the blue component to bright blue. If the col or parameter has a high value, as it does for the control that appears to the left, the red and green components are fairly bright, too, giving a whitish-blue color. For controls towards the right, where the moisture is higher, we dim the red and green color to make the controls a deeper blue.

```
! Create the moisture bar
DIM moisture AS Progress
moisture = Graphics.newProgress(20, 165, width - 40)
```

The last control is the progress bar used as an analog moisture meter.

```
! Set HiJack to sample 10 times
! per second
HiJack.setRate(10)
```

The HiJack hardware has a variable sample rate. We can leave it at the default of about 42 Hz, but higher sampling rates use more power. Lowering it to 10 Hz saves power, and let's face it, even 10 samples per second is overkill for a moisture meter.

So far, we've done everything except actually read the HiJack sensor and display the results. We want to do that on a regular basis for as long as the program runs. To do that in an event-driven program, we create a nullEvent subroutine. It will be called repeatedly when the program is not doing something else, which at this point means it will be called all of the time.

```
! Read and process HiJack values
!
! Parameters:
!    time - Event time

SUB nullEvent (time AS DOUBLE)
v = HiJack.receive ❶
m = -5.385531 + 0.07708497*v(1) ❷
IF m < 0 THEN m = 0 ❸
IF m > 4 THEN m = 4
moisture.setValue(m/4) ❹
value.setText(STR(INT(m*10)/10))
END SUB
```

Here's what's happening in the nullEvent subroutine:

❶ Each time this subroutine is called, it starts by calling `HiJack.receive`. This fetches a two-element array from the HiJack sensor. The first element is the 0–255 value returned by the A–D converter, while the second is a timestamp indicating when the data was collected.

❷ The next line converts the HiJack value to a moisture reading from 0–4, using the fit we got from calibrating the moisture meter earlier.

❸ The following two lines pin the value to the desired range, discarding values that are above or below the expected range.

❹ Finally, we set the analog moisture meter—dividing by 4 because progress bars expect a value between 0 and 1—and the digital readout, using the `INT` function to strip off all but one decimal point from the result.

The Complete Moisture Meter Source

Here's the complete source for the moisture meter program. Don't type this in! It's just here for your reference while reading. The complete program is in the *O'Reilly Books* folder in techBASIC and techBASIC Sampler. Look for the program called Moisture Meter.

```
! HiJack Moisture Meter

! Show the graphics screen
System.showGraphics(1)
System.setAllowedOrientations(1)
Graphics.setToolsHidden(1)

! Get the size of the graphics
! screen
width = Graphics.width
height = Graphics.height

! Paint the background light gray
bg = 0.9
Graphics.setColor(bg, bg, bg)
Graphics.fillRect(0, 0, width, height)

! Create a Quit button
DIM quit AS Button
quit = Graphics.newButton(width/2 - 36, height - 57)
quit.setTitle("Quit")
quit.setBackgroundColor(1, 1, 1)
quit.setGradientColor(0.7, 0.7, 0.7)

! Put the name of the program at
! the top of the screen
DIM mmLabel AS Label
mmLabel = newLabel(0, 20, width, 40, 40, "Moisture Meter")
```

```
mmLabel.setBackgroundColor(bg, bg, bg)

! Create a large label to show
! the moisture level
DIM value AS Label
value = newLabel(0, 75, width, 40, 50, "0")

! Add 5 small labels to show the
! moisture scale along the top of
! the moisture bar
DIM nums(5) AS Label
plantLabelWidth = (width - 40)/4
FOR i = 0 TO 4
  x = i*plantLabelWidth
  nums(i + 1) = newLabel(x, 140, 40, 20, 16, STR(i))
NEXT

! Create the strings that will
! name the plants in each
! moisture group
DIM plants(4) AS TextView, plants$(4)
addPlant("Aloe", plants$(1))
addPlant("Geranium", plants$(1))
addPlant("Jade Plant", plants$(1))
addPlant("Orchid", plants$(1))
addPlant("Wandering Jew", plants$(1))
addPlant("African Violet", plants$(2))
addPlant("Cacti", plants$(2))
addPlant("Hibiscus", plants$(2))
addPlant("Wax Plant", plants$(2))
addPlant("Begonia", plants$(3))
addPlant("Flowering Maple", plants$(3))
addPlant("Peppers", plants$(3))
addPlant("Spider Plant", plants$(3))
addPlant("Azalea", plants$(4))
addPlant("Ferns", plants$(4))
addPlant("Melons", plants$(4))
addPlant("Peace Lily", plants$(4))
addPlant("Tomatoes", plants$(4))

! Add colored labels below the
! moisture bar showing the plants
! in each group
plantLabelHeight = 150
FOR i = 1 TO 4
  x = 20 + (i - 1)*plantLabelWidth
  color = 1 - i/5
  plants(i) = newTextView(x, 170, plantLabelWidth, plantLabelHeight, <?pdf-cr?>
  11, color, plants$(i))
NEXT

! Create the moisture bar
```

```
DIM moisture AS Progress
moisture = Graphics.newProgress(20, 165, width - 40)

! Set HiJack to sample 10 times
! per second
HiJack.setRate(10)

! Create a label
!
! Parameters:
!   x - Horizontal location
!   y - Vertical location
!   width - Label width
!   height - Label height
!   fontSize - Point size for the
!      font
!   text$ - Label text
!
! Returns: The label

FUNCTION newLabel (x, y, width, height, fontSize, text$) AS Label
DIM nl AS Label
nl = Graphics.newLabel(x, y, width, height)
nl.setText(text$)
nl.setBackgroundColor(1, 1, 1, 0)
nl.setAlignment(2)
nl.setFont("Sans_Serif", fontSize, 0)
newLabel = nl
END FUNCTION

! Add a plant name to a string
! containing plant names
!
! Parameters:
!   newPlant$ - New plant name
!   plant$ - Current plant names

SUB addPlant (newPlant$, BYREF plant$)
IF LEN(plant$) <> 0 THEN
  plant$ = plant$ & CHR(10) & CHR(10)
END IF
plant$ = plant$ & newPlant$
END SUB

! Create a text view to show a
! list of plants
!
! Parameters:
!   x - Horizontal location
!   y - Vertical location
!   width - TextView width
!   height - TextView height
```

```
!   fontSize - Point size for the
!       font
!   color - White level for
!       background; the color will
!       be blue, lightened by this
!       amount
!   text$ - TextView text
!
! Returns: The text view

FUNCTION newTextView (x, y, width, height, fontSize, color, text$) AS TextView
DIM ntv AS TextView
ntv = Graphics.newTextView(x, y, width, height)
ntv.setText(text$)
ntv.setEditable(0)
ntv.setBackgroundColor(color, color, 1, 1)
IF color < 0.5 THEN
  ntv.setColor(1, 1, 1)
END IF
ntv.setAlignment(2)
ntv.setFont("Sans_Serif", fontSize, 0)
newTextView = ntv
END FUNCTION


! Handle a tap on a button
!
! Parameters:
!   ctrl - The button tapped
!   time - When the button was
!       tapped

SUB touchUpInside(ctrl AS Button, time AS DOUBLE)
IF ctrl = quit THEN
  STOP
END IF
END SUB


! Read and process HiJack values
!
! Parameters:
!    time - Event time

SUB nullEvent (time AS DOUBLE)
v = HiJack.receive
m = -5.385531 + 0.07708497*v(1)
IF m < 0 THEN m = 0
IF m > 4 THEN m = 4
moisture.setValue(m/4)
value.setText(STR(INT(m*10)/10))
END SUB
```

CHAPTER 6

Bluetooth Low Energy

About This Chapter

Prerequisites

Read Chapter 1 and the end of Chapter 2 (the section on the techBASIC help system) if you need some help with techBASIC.

Equipment

You will need an iPhone 4S or later, iPod 5th Gen or later, or iPad 3 or later running iOS 5 or later. The chapter uses the Texas Instruments SensorTag as a sample device. You could follow along with any Bluetooth low energy device, but obviously it will be simpler to use the SensorTag. See Table 6-1 for a complete parts list.

Software

You will need a copy of techBASIC or techBASIC Sampler.

This chapter introduces Bluetooth low energy, a rapidly expanding technology for sensors. It's used in the next two chapters for accessing sensor data from a model rocket and for controlling an RC truck using an Arduino microcontroller, just two of the many ways Bluetooth low energy technology can be used from an iOS device.

Chapters 9 and 10 show how to convert an iOS device so it becomes a Bluetooth low energy device that can be sensed and read by other devices, including other iOS devices.

What Is Bluetooth Low Energy?

Bluetooth low energy, also known as Bluetooth 4.0, BLE, Bluetooth LE, and Bluetooth Smart, is a new twist on the old Bluetooth standard. It's designed for ultra-low-energy sensors that can run off of coin cell batteries. Some of the Bluetooth low energy devices that are already available include sports fitness sensors like bicycle speedometers and pedometers built into Nike running shoes; health sensors like thermometers, blood pressure cuffs, and heart rate monitors; and physics sensors like accelerometers and thermometers.

Keep in mind that Bluetooth and Bluetooth low energy are two entirely different technologies whose main commonality is that they share a name. Both use radio signals to communicate between devices, but the radio signals are not compatible. A Bluetooth low energy device can't communicate with a Bluetooth controller, or vice versa. Occasionally you may see a reference to a dual-mode device; this is a single physical device that has both Bluetooth and Bluetooth low energy components built in, so they can communicate either way, but that's the exception, not the rule.

So why have two standards? Well, they are actually designed for very different applications. Bluetooth is optimized for fast, relatively high-power communication between devices that are semipermanently paired with one another. Typical examples are keyboards (paired with a computer) and ear buds (paired with a cell phone). It's not uncommon to have several people talking near one another using Bluetooth ear buds, and Bluetooth facilitates this nicely by making sure one specific ear bud always talks to one specific cell phone.

In contrast, a Bluetooth low energy device is set up for communication with just about any device that wants to talk to it. The devices are called *slaves*; they broadcast a "here I am" signal called an *advertisement* on a fairly constant basis. A Bluetooth low energy based heart rate monitor is constantly sending out a signal, for example. *Master* devices, like iPhones, listen for these advertisements and either gather information right from the advertisement itself or connect with the device to send information back and forth.

Bluetooth low energy devices are also optimized for very low energy usage, typically running for weeks or months on a single coin cell battery. To keep the energy use low, radio signals are kept very short. Bluetooth low energy is not the technology to use for devices that generate high volumes of traffic, like security cameras, but it's great for devices that require simple, infrequent communication, like an alarm on a door or a heart rate monitor that will send out a couple of bytes indicating a pulse rate every few seconds.

There is another huge difference between Bluetooth and Bluetooth low energy for iPhone and iPad enthusiasts. Bluetooth access is covered by Apple's MFi program, which requires anyone creating a Bluetooth device to buy and use a special Apple chip. Apple routinely rejects all apps that use a Bluetooth device except for the original app developed by the vendor for the hardware. Even if you know exactly how to connect to a Bluetooth device and have approval from the device's creator, Apple will almost certainly reject your Bluetooth app if you submit it to the App Store. On the other hand, the MFi program does not cover Bluetooth low energy devices or apps. Apps that support

Bluetooth low energy devices are routinely accepted for distribution on the App Store. In fact, the program described in this chapter is available on the App Store; it's called SensorTag.

The TI SensorTag

This chapter explores Bluetooth low energy using the Texas Instruments SensorTag. The SensorTag is designed to help hardware and software engineers learn about and begin building Bluetooth low energy peripherals and software using the CC2541 chip. It has these six sensors:

- A three-axis accelerometer
- A three-axis magnetometer
- A three-axis gyroscope
- A dual-mode thermometer that measures ambient and remote temperatures
- A humidity sensor
- A barometer

We will use each of these six sensors to build the app shown in Figure 6-1, and in the process, we'll learn about the concepts used to connect to absolutely any Bluetooth low energy device.

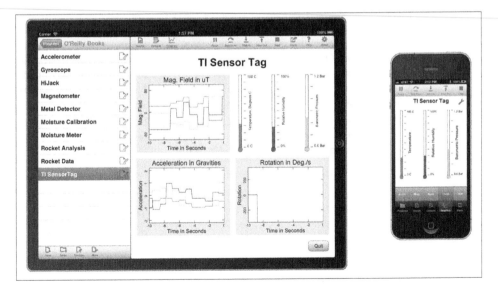

Figure 6-1. The SensorTag app

While it will help to have the SensorTag device as you work through the examples in this chapter, you can follow along with any Bluetooth low energy device. The two chapters that follow will put all of this to use, first to gather information from a rocket flight using Bluetooth low energy, then to hack a radio-controlled car to steer it from an iPhone using Bluetooth low energy and an Arduino microcontroller.

Table 6-1 lists the parts required to build the SensorTag app, all of which are shown in Figure 6-2.

Table 6-1. Parts list for SensorTag app

Part	Description
iPhone 4S or later, iPad 3 or later, or iPod Touch 5th Gen or later	Bluetooth low energy devices are physically different from the older Bluetooth devices. Bluetooth low energy requires completely new hardware that is not available in older model devices, so you will need an iPhone 4s or later, an iPad 3 or later, or an iPod Touch 5th Gen (the one with the 4″ retina display) or later to access Bluetooth low energy devices.
TI SensorTag	The SensorTag is a Bluetooth low energy device with six sensors. It's physically rather small, with the slightly irregularly shaped soft plastic case measuring about $2\frac{3}{4}$″ x $1\frac{1}{2}$″ x $\frac{1}{2}$″. The circuit board is 1″ x $2\frac{1}{4}$″ x $\frac{1}{4}$″. It runs on a single CR 2032 coin cell battery, and mine has been chugging along quite nicely for several weeks on the same battery. It's available directly from Texas Instruments (*http://www.ti.com/sensortag*). A great place for additional information is the SensorTag wiki (*http://www.ti.com/sensortag-wiki*).
CC Debugger	The CC Debugger is used to load new firmware onto the SensorTag, as well as to connect diagnostic software. The SensorTag comes with a basic set of firmware preinstalled, so technically this is optional, but you may need to upgrade the firmware eventually. For example, in the next chapter we will change the firmware to switch the accelerometer from ±2G to ±8G for our rocket flights. You may already have a CC Debugger from the older Texas Instruments Bluetooth Low Energy CC2540 Mini Development Kit, which included the Key Fob. If not, I'd recommend getting one when you order the SensorTag. Like the SensorTag, it's available directly from Texas Instruments. The software for the CC Debugger runs on a Windows operating system. You only need it when you install the firmware, but if you're a Mac person, plan ahead so you have access to a Windows computer when you need it. We don't need to install firmware for this chapter, so the techniques won't be discussed here. See "SensorTag 8G Software" on page 185 in Chapter 7 for instructions on installing firmware.
techBASIC	Bluetooth low energy support is built into techBASIC starting with version 2.3. This is available as a free update to people who own older versions of techBASIC. New copies can be purchased in the App Store (*http://itunes.apple.com/us/app/techbasic/id470781862?ls=1&mt=8*).

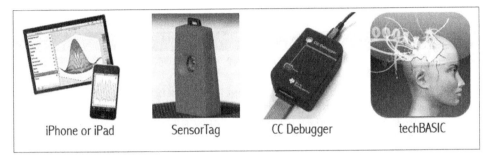

| iPhone or iPad | SensorTag | CC Debugger | techBASIC |

Figure 6-2. Photos of SensorTag and CC Debugger, courtesy of Jarle Bøe, Texas Instruments

Writing Bluetooth Low Energy Programs

With the hardware in hand, it's time to write the software. The next few pages develop a short text program to access the accelerometer on the SensorTag. The point is to show how to access Bluetooth low energy programs, so it's kept simple and uses text output. Later in the chapter, we'll develop more sophisticated programs to access each of the six sensors, going over each in detail. These are text programs, too. The program shown over the next few pages only appears here in the book. The programs for the six individual sensor programs that follow are all based on this one, though, and all six of those can be found in the *O'Reilly Books* folder in either techBASIC or techBASIC Sampler—so don't start typing what you see here unless that's the way you learn (it is for me!).

There is also a nice GUI program to access the SensorTag, shown in Figure 6-1. It shows the acceleration, rotation, and magnetic field on interactive plots, and displays the temperature, humidity, and barometric pressure on stylized thermometers. It works on both the iPhone and iPad. It's called SensorTag and is available as a free download from the App Store (*https://itunes.apple.com/us/app/sensortag/id579408063?ls=1&mt=8*). The source code for SensorTag can also be found in the *O'Reilly Books* folder of either techBASIC or techBASIC Sampler. Table 6-2 lists the various apps seen in this chapter, some free-standard and some with source for the programs you will create and run.

Table 6-2. SensorTag apps

App	Description
SensorTag	This free app is available for download from the App Store. This is a compiled version of the program from this chapter running as a standalone app.
techBASIC Sampler	This is the demo version of techBASIC. It's a free app. The source code for the GUI app shown in this chapter is in the folder *O'Reilly Books*; the program is called TI SensorTag. The six text-based SensorTag programs from this chapter are in the same folder. You can view and run the program from this version of techBASIC, but you won't be able to edit the program or create new ones.
techBASIC	The full version of techBASIC is also available from the App Store. As with techBASIC Sampler, the source for the SensorTag programs is in the *O'Reilly Books* folder.

Before jumping into the code, though, let's stop and get an overview of how Bluetooth low energy devices are designed. Bluetooth low energy devices package information in *services*. In our program, we will use six of these services, one for each sensor. Other devices might offer a heart rate service or a servo control service.

Each service has zero or more *characteristics*, which work more or less like variables. Some can only be read, some can only be written, and some can be both read and written. There can be more than one characteristic for a service. For example, the accelerometer has three characteristics: one to start or stop the accelerometer, one to set the rate at which it will return data, and one to read the actual data. Services can also have other subservices embedded in them; these are called *included services*.

Characteristics can also have *descriptors* with additional information about the characteristic.

Figure 6-3 shows the services and characteristics we'll use on the SensorTag. This is a simplification of the complete specification for the communication protocol for the SensorTag. The full specification is called the *GATT profile* and is available for download from the TI website (*http://www.ti.com/sensortag*). There is also a lot of supplemental information on each sensor in the wiki (*http://www.ti.com/sensortag-wiki*). You don't need the GATT profile or the wiki to follow along, but learning to read these documents will help if you have to figure out another Bluetooth low energy device.

Almost all Bluetooth low energy calls are asynchronous, since it may take some time for the operating system to communicate with the device to carry out an operation. Your program can use the time waiting for the results of a call to make new requests, handle information passed back from old requests, or simply do something else. From the program's perspective, the program begins by making a call, then moves on. At some point in the future the operating system will call a subroutine in the program to report the results. You will see this pattern over and over in the SensorTag program.

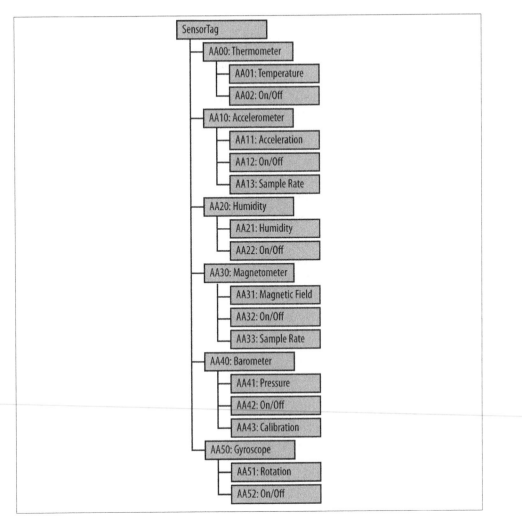

Figure 6-3. SensorTag services

Let's get going with the program. The first step in connecting to a Bluetooth low energy device is to start the Bluetooth low energy service with the call:

```
BLE.startBLE
```

Next, we begin scanning for devices. This allows our program to look for any Bluetooth low energy devices in the area and connect to the one—or ones—with the information or capabilities we want. In general, you should already know the kind of service you are looking for. Each service has a 16- or 128-bit identifier, called the service *UUID* (Universal Unique Identifier). The shorter 16-bit identifiers are supposed to be assigned by the Bluetooth standards committee. You can find a list of the standard services on the

Bluetooth website (*http://developer.bluetooth.org/gatt/services/Pages/Service sHome.aspx*). Anyone is free to create a service using a 128-bit UUID.

The values shown in Figure 6-3 are actually only part of the 128-bit UUIDs for the characteristics and services. These values each make up the third and fourth bytes (fifth through eighth hexadecimal digits) of a much longer UUID. The complete UUID for the accelerometer service looks like this:

```
F000AA00-0451-4000-B000-000000000000
```

The UUID strings for other services and characteristics are formed by replacing the third and fourth bytes (fifth through eighth digits, AA00 in this case) with the appropriate values from the table, so the calibration UUID for the barometer looks like this:

```
F000AA43-0451-4000-B000-000000000000
```

Replacing just four digits to distinguish between services is not a universal pattern with UUIDs, just a convention Texas Instruments followed for this device that makes it a bit easier to deal with the UUIDs for the various services.

In our case, we're going to scan for any Bluetooth low energy peripheral in the area by passing an empty array of UUIDs to the startScan method:

```
DIM uuid(0) AS STRING
BLE.startScan(uuid)
```

At this point, the iPhone or iPad starts looking around for any Bluetooth low energy devices in the area. Bluetooth low energy devices advertise their presence with short, infrequent radio signals that our program is scanning for. As the iOS device finds Bluetooth low energy devices, it calls a subroutine in the program called BLEDiscovered Peripheral. Here's the implementation from our program:

```
! Set up variables to hold the peripheral and the characteristics
! for the battery and buzzer.
DIM sensorTag AS BLEPeripheral

! Start the BLE service and begin scanning for devices.
debug = 1
BLE.startBLE
DIM uuid(0) AS STRING
BLE.startScan(uuid)

! Called when a peripheral is found. If it is a Sensor Tag, we
! initiate a connection to it and stop scanning for peripherals.
!
! Parameters:
!    time - The time when the peripheral was discovered.
!    peripheral - The peripheral that was discovered.
!    services - List of services offered by the device.
!    advertisements - Advertisements (information provided by the
!        device without the need to read a service/characteristic)
```

```
!    rssi - Received Signal Strength Indicator
!
SUB BLEDiscoveredPeripheral (time AS DOUBLE, _
                            peripheral AS BLEPeripheral, _
                            services() AS STRING, _
                            advertisements(,) AS STRING, _
                            rssi)
IF peripheral.bleName = "TI BLE Sensor Tag" THEN
   sensorTag = peripheral
   BLE.connect(sensorTag)
   BLE.stopScan
   IF debug THEN PRINT "Discovered SensorTag."
END IF
END SUB
```

There are several ways to pick from the Bluetooth low energy devices that may be out there. One is to scan for a particular kind of service, say a thermometer, and use the first one we find. Because in this case we're more interested in a kind of device rather than a kind of service, our subroutine checks to see if we've found the device we're looking for by looking at the name of the peripheral the iPhone or iPad has found. If it matches TI BLE Sensor Tag, we've found what we are looking for. Ideally, things like the name of the peripheral and the peripheral UUID will be in the documentation, but in practice, you may have to write just this much of the program and print the name of any peripheral you find to figure out the name to use.

Once we find the device, we save the peripheral in a global variable. This is important —it keeps the memory manager from disposing of the peripheral's object when the subroutine ends, which would tell the operating system we're not interested in this peripheral. Next, we attempt to connect to the peripheral using BLE.connect. The last step is to stop scanning for other Bluetooth low energy devices, which can drain the battery of the iOS device and the Bluetooth low energy devices. We do this with BLE.stopScan.

This tiny program is actually a complete working program to find and connect to a SensorTag. To see it in action, run the program, then press the pairing button on the left side of the SensorTag (shown in Figure 6-4).

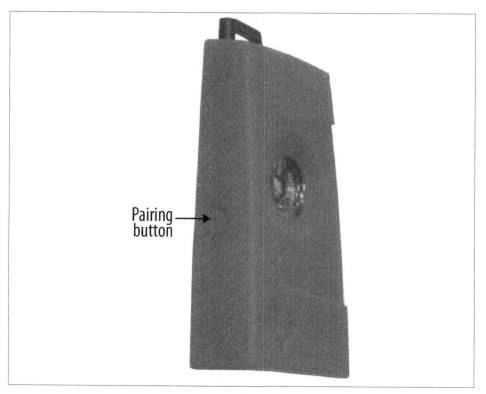

Figure 6-4. SensorTag pairing button

It will take a second or two for the iPhone and SensorTag to set up communications, then you should see:

```
Discovered SensorTag.
```

printed to the console.

Debugging Connection Issues

Most of the time a connection just works. That's what is supposed to happen. What if it doesn't?

There are all sorts of reasons a connection might fail. Here are the common ones, along with things you can try to correct the problem:

- The battery might be dead. It happens. Sometimes the device itself will have a power indicator. If not, try changing the battery.
- The device may not be advertising. Some devices advertise more or less constantly, while others, like the SensorTag, need to be prompted to do so using a pairing button. Check the device, and press the pairing button if needed.
- There may be a lot of radio interference from other sources. It's rare, but the possibility exists.
- The software may not be working. For the programs in this book, there are two common reasons.
  - The first is that the program has been typed incorrectly.
  - The second is more interesting. What if the name of the device is not `TI BLE Sensor Tag`? That could easily happen due to a change in the firmware on the device itself. The sniffer program mentioned earlier is a great way to check.
- Occasionally, a device just doesn't seem to want to connect, or connects very slowly. This really should not happen, and it may not happen to you as iOS and techBASIC change. I have seen this with multiple apps and multiple devices, though, so it does not appear to be a problem with the SensorTag or with techBASIC. It does happen more often when multiple apps are used to access the same device. If this happens, try the following, in this order:
  - Shut techBASIC down completely and restart it by double-tapping the home button, pressing the techBASIC icon until it jiggles, and tapping the red – icon to close the program. Tap the home button to return to the home screen, then restart techBASIC. This probably won't work, but it's the easiest thing to try.
  - Turn Bluetooth off and back on from the Settings app. This works most of the time.
  - In really stubborn cases, restart the device. In every case I have seen where the device was working and had fresh batteries, this has always cured the connection problem.

Let's pause and consider what this program is doing. Our ultimate goal is to connect to the TI SensorTag, but this program is looking for literally any Bluetooth low energy device it can find. Change the subroutine to:

```
SUB BLEDiscoveredPeripheral (time AS DOUBLE, _
                             peripheral AS BLEPeripheral, _
                             services() AS STRING, _
                             advertisements(,) AS STRING, _
                             rssi)
    PRINT "Found "; peripheral.uuid; _
        " with a name of "; peripheral.bleName
END SUB
```

and you have a simple Bluetooth low energy sniffer program that will constantly write the UUIDs and names of any Bluetooth low energy peripherals it finds to the console. It's a simple program, but it's also a complete, useful sniffer that will help identify new Bluetooth low energy devices in the area.

Here's an example of what it will print for a SensorTag:

```
Found 00000000-0000-0000-C47F-FD159A33F73C with a name
of TI BLE Sensor Tag
```

Connecting to the device does not happen right away. The operating system asks for access and, once it gets a response, calls another subroutine called BLEPeripheralInfo:

```
! Called to report information about the connection status of the
! peripheral or to report that services have been discovered.
!
! Parameters:
!     time - The time when the information was received.
!     peripheral - The peripheral.
!     kind - The kind of call. One of
!         1 - Connection completed
!         2 - Connection failed
!         3 - Connection lost
!         4 - Services discovered
!     message - For errors, a human-readable error message.
!     err - If there was an error, the Apple error number. If there
!         was no error, this value is 0.
!
SUB BLEPeripheralInfo (time AS DOUBLE, _
                       peripheral AS BLEPeripheral, _
                       kind AS INTEGER, _
                       message AS STRING, _
                       err AS LONG)
    IF kind = 1 THEN
```

```
    ! The connection was established. Look for available services.
    IF debug THEN PRINT "Connection made."
    peripheral.discoverServices(uuid)
ELSE IF kind = 2 OR kind = 3 THEN
    IF debug THEN PRINT "Connection lost: "; kind
    BLE.connect(sensorTag)
ELSE IF kind = 4 THEN
    ! Services were found. If it is one of the ones we are interested
    ! in, begin discovery of its characteristics.
    DIM availableServices(1) AS BLEService
    availableServices = peripheral.services
    FOR s = 1 to UBOUND(services, 1)
      FOR a = 1 TO UBOUND(availableServices, 1)
        IF services(s) = availableServices(a).uuid THEN
          IF debug THEN PRINT "Discovering characteristics for "; services(s)
          peripheral.discoverCharacteristics(uuid, availableServices(a))
        END IF
      NEXT
    NEXT
END IF
END SUB
```

We also need to add these lines at the start of the program, right after DIM sensorTag AS BLEPeripheral:

```
! We will look for these services.
DIM servicesHeader AS STRING, services(1) AS STRING
servicesHeader = "-0451-4000-B000-000000000000"
services(1) = "F000AA10" & servicesHeader : ! Accelerometer
accel% = 1
```

This subroutine can get called for a variety of reasons. After a BLE.connect call, we can get back the "connection complete" response or be told that the connection failed. Later, the connection might be lost—perhaps we've wandered too far away from the peripheral. The kind parameter tells us why the call was made. Our program asks the peripheral for a list of any services it provides by calling the peripheral's discoverServices method when a connection is made, and tries to reconnect if a connection is lost. As the peripheral reports back on any services, the subroutine is called again with kind set to 4. Since we've enabled the debug output, the program will print a list of the available services.

Eventually we'll expand the array to hold all six services, but for now, we'll restrict our attention to the accelerometer, so the array only has one element.

The code in BLEPeripheralInfo checks to see if the service reported by the device is the accelerometer and, if so, asks the service for a list of the available characteristics using the peripheral's discoverCharacteristics method.

The Bluetooth Low Energy Sniffer, Part 2

Continuing with our sniffer, this implementation of BLEPeripheralInfo asks any device it finds for a list of available services and prints the services. This will work with any Bluetooth low energy device:

```
! Set up variables to hold the peripheral and the characteristics
DIM sensorTag AS BLEPeripheral

BLE.startBLE
DIM uuid(0) AS STRING
BLE.startScan(uuid)


SUB BLEDiscoveredPeripheral (time AS DOUBLE, _
                            peripheral AS BLEPeripheral, _
                            services() AS STRING, _
                            advertisements(,) AS STRING, _
                            rssi)
BLE.connect(peripheral)
END SUB


SUB BLEPeripheralInfo (time AS DOUBLE, _
                       peripheral AS BLEPeripheral, _
                       kind AS INTEGER, _
                       message AS STRING, _
                       err AS LONG)
IF kind = 1 THEN
  peripheral.discoverServices(uuid)
ELSE IF kind = 4 THEN
  DIM services(1) AS BLEService
  services = peripheral.services
  FOR s = 1 to UBOUND(services, 1)
    PRINT "Found service. Peripheral = "; peripheral.uuid
    PRINT "               Service = "; services(s).uuid
    PRINT
  NEXT
END IF
END SUB
```

Here's the output when it discovered a SensorTag:

```
Found service. Peripheral = 00000000-0000-0000-C47F-FD159A33F73C
               Service = 1800

Found service. Peripheral = 00000000-0000-0000-C47F-FD159A33F73C
```

```
                       Service = 1801

    Found service. Peripheral = 00000000-0000-0000-C47F-FD159A33F73C
                       Service = 180A

    Found service. Peripheral = 00000000-0000-0000-C47F-FD159A33F73C
                       Service = F000AA00-0451-4000-B000-000000000000

    Found service. Peripheral = 00000000-0000-0000-C47F-FD159A33F73C
                       Service = F000AA10-0451-4000-B000-000000000000

    Found service. Peripheral = 00000000-0000-0000-C47F-FD159A33F73C
                       Service = F000AA20-0451-4000-B000-000000000000

    Found service. Peripheral = 00000000-0000-0000-C47F-FD159A33F73C
                       Service = F000AA30-0451-4000-B000-000000000000

    Found service. Peripheral = 00000000-0000-0000-C47F-FD159A33F73C
                       Service = F000AA40-0451-4000-B000-000000000000

    Found service. Peripheral = 00000000-0000-0000-C47F-FD159A33F73C
                       Service = F000AA50-0451-4000-B000-000000000000

    Found service. Peripheral = 00000000-0000-0000-C47F-FD159A33F73C
                       Service = FFE0

    Found service. Peripheral = 00000000-0000-0000-C47F-FD159A33F73C
                       Service = F000AA60-0451-4000-B000-000000000000
```

By this time, you probably have a good idea what will happen next. As with the services, the characteristics are reported to the program by calling a subroutine. In this case, the subroutine is BLEServiceInfo. The name might seem odd, but the information is about a service, not the characteristic. The operating system is telling us the service has information. Here's the implementation:

```
! Called to report information about a characteristic or included
! services for a service. If it is one we are interested in, start
! handling it.
!
! Parameters:
!    time - The time when the information was received.
!    peripheral - The peripheral.
!    service - The service whose characteristic or included
!        service was found.
!    kind - The kind of call. One of
!        1 - Characteristics found
!        2 - Included services found
!    message - For errors, a human-readable error message.
!    err - If there was an error, the Apple error number. If there
!        was no error, this value is 0.
```

```
    !
    SUB BLEServiceInfo (time AS DOUBLE, _
                        peripheral AS BLEPeripheral, _
                        service AS BLEService, _
                        kind AS INTEGER, _
                        message AS STRING, _
                        err AS LONG)
    IF kind = 1 THEN
      ! Get the characteristics.
      DIM characteristics(1) AS BLECharacteristic
      characteristics = service.characteristics
      FOR i = 1 TO UBOUND(characteristics, 1)
        IF service.uuid = services(accel%) THEN
          ! Found the accelerometer.
          SELECT CASE characteristics(i).uuid
            CASE "F000AA11" & servicesHeader
              ! Tell the accelerometer to begin sending data.
              IF debug THEN PRINT "Start accelerometer."
              DIM value(2) as INTEGER
              value = [0, 1]
              peripheral.writeCharacteristic(characteristics(i), value, 0)
              peripheral.setNotify(characteristics(i), 1)

            CASE "F000AA12" & servicesHeader
              ! Turn the accelerometer sensor on.
              IF debug THEN PRINT "Accelerometer on."
              DIM value(1) as INTEGER
              value(1) = 1
              peripheral.writeCharacteristic(characteristics(i), value, 1)

            CASE "F000AA13" & servicesHeader
              ! Set the sample rate to 100ms.
              DIM value(1) as INTEGER
              value(1) = 100
              IF debug THEN PRINT "Setting accelerometer sample rate to "; value(1)
              peripheral.writeCharacteristic(characteristics(i), value, 1)
          END SELECT
        END IF
      NEXT
    END IF
    END SUB
```

We're only interested in the first kind of call, where the operating system is telling us the service has updated its list of characteristics. If kind is 1, we get the characteristics for the service with a call to the service's characteristics method and loop over them, looking for characteristics we're interested in. There's only one so far, but the list will grow in the final program.

But wait—it looks like the characteristics are part of the service, which we knew after the call to BLEPeripheralInfo. Why go to all this trouble? The reason is that the operating system doesn't ask the device for a list of characteristics until the discoverChar acteristics call, since it doesn't want to waste battery power asking for information unless it is really needed. You can call the service's characteristics method in BLE PeripheralInfo, but it will return an empty array.

There is more than one way to read a value from the device. The accelerometer uses a method called *notification*, where the device notifies the iPhone each time a new value is available. We turn notifications on and start them with the calls:

```
DIM value(2) as INTEGER
value = [0, 1]
peripheral.writeCharacteristic(characteristics(i), value, 0)
peripheral.setNotify(characteristics(i), 1)
```

The first two lines set up the value to write to the peripheral. The accelerometer expects two bytes, a 0 and a 1, to turn on notifications. The next line writes these values to the characteristic, telling it to report information using notifications. The last line actually starts the notifications.

The other method would look similar. To read the accelerometer a single time, you could use the readCharacteristics call, like this:

```
peripheral.readCharacteristic(characteristics(i))
```

Most characteristics support either notification or individual reads, but not both. As it turns out, the accelerometer does support both types of calls. We'll see later how to choose between them.

The BLECharacteristicInfo subroutine gets called with either notifications or reads, the difference being that it will only be called once after a read but will be called every time a new value is available for notifications.

The next two characteristics are used to turn the accelerometer on and set the sampling rate. They work pretty much like the write used earlier to tell the device to report information using notifications. It may seem odd to send information to the command using an array with only one element, but that's because the software making the call to the device really has no idea how many bytes to send, so it sends all of the bytes in any array passed. In these cases, the device only expects a single byte. For the AA12 characteristic, a 1 turns the accelerometer on and a 0 turns it off. For the AA13 characteristic, the value is the number of tens of milliseconds between samples, so passing 100 tells the device to send back an acceleration once a second.

The operating system calls the BLECharacteristicInfo subroutine when the device reports a change to a characteristic. Here's a simplified version of the subroutine that appears in the SensorTag app. This one just handles the accelerometer, while the SensorTag app deals with all six sensors in the equivalent subroutine:

```
! Called to return information from a characteristic.
!
```

```
! Parameters:
!    time - The time when the information was received.
!    peripheral - The peripheral.
!    characteristic - The characteristic whose information
!        changed.
!    kind - The kind of call. One of
!        1 - Called after a discoverDescriptors call.
!        2 - Called after a readCharacteristics call.
!        3 - Called to report status after a writeCharacteristics
!            call.
!    message - For errors, a human-readable error message.
!    err - If there was an error, the Apple error number. If there
!        was no error, this value is 0.
!
SUB BLECharacteristicInfo (time AS DOUBLE, _
                            peripheral AS BLEPeripheral, _
                            characteristic AS BLECharacteristic, _
                            kind AS INTEGER, _
                            message AS STRING, _
                            err AS LONG)
IF kind = 2 THEN
  DIM value(1) AS INTEGER
  value = characteristic.value
  SELECT CASE characteristic.uuid
    CASE "F000AA11" & servicesHeader
      ! Update the accelerometer.
      c = 64.0
      p% = value(1)
      IF p% BITAND $0080 THEN p% = p% BITOR $FF00
      lastAccelX = p%/c

      p% = value(2)
      IF p% BITAND $0080 THEN p% = p% BITOR $FF00
      lastAccelY = p%/c

      p% = value(3)
      IF p% BITAND $0080 THEN p% = p% BITOR $FF00
      lastAccelZ = p%/c

      PRINT lastAccelX, lastAccelY, lastAccelZ

  END SELECT
ELSE IF kind = 3 AND err <> 0 THEN
  PRINT "Error writing "; characteristic.uuid; ": ("; err; ") "; message
END IF
END SUB
```

This subroutine is called with a kind of 2 when the accelerometer reports back with a new value. The value it returns is packed into a three-byte array, where each byte is the acceleration value along one axis. These are signed values with a range of –128 to 127, so we can't just use them as is. Instead, the program checks to see if the value is negative and, if so, extends the sign bits to form a two-byte signed integer.

Bits and Bytes and Words, Oh My!

These lines:

```
p% = value(1)
IF p% BITAND $0080 THEN p% = p% BITOR $FF00
```

may seem a bit odd, even to an experienced BASIC programmer. They are doing some tricks to convert bytes to integer values. Here is a very brief description of what is happening. See the two's complement Wikipedia article (*http://en.wikipedia.org/wiki/Two %27s_complement*) or any good introductory computer science book for a more complete discussion.

The bytes that come back from the Bluetooth low energy device are made up of eight bits. By convention, these eight bits are usually used to represent the values 0 to 255 using binary arithmetic. They can also represent the values –128 to 127 using two's complement notation, which is what is happening in this case. With two's complement notation, negative values have their most significant bit set—the bit $0080 for a single byte. p% BITAND $0080 will be nonzero if that bit is set, and zero for positive numbers.

If the bit is set, P% BITOR $FF00 is used to *sign extend* the shorter byte value so it fills the full two bytes used for an integer in techBASIC. This creates a number that represents the same value as the original byte, but does so in a longer integer value.

There will be several other places in the book where bits and bytes will be manipulated to convert from one data type to another or to extract information that is coded as something other than a standard number. We'll go over each case in detail. This will be old hat if you are already familiar with how bits and bytes are used in low-level communications. If it's new to you, be sure and work through the examples. Understanding how bits and bytes can be used to code information, and how to pack and unpack the information from a high-level language, is an essential part of dealing with sensors and other hardware.

Run the program. (It's available in techBASIC and techBASIC Sampler as SensorTag Accelerometer in the *O'Reilly Books* folder.) It may connect to the peripheral right away if you have already run an earlier version of the program. If not, push the pairing button on the SensorTag to establish a connection. After a second or two, you should start to see a spew of accelerometer data on the console telling you the G forces experienced by the SensorTag. Twist it around slowly to verify that the value changes. Here are the first few lines I saw:

```
0            -0.9375      0.234375
0.015625     -0.875       0.453125
```

```
0.46875        -0.421875      0.765625
0.46875        0              0.953125
0.65625        0.71875        0.40625
0.640625       0.6875         0.265625
```

The SensorTag coordinate system, like the iPhone's coordinate system, defines how the sensors are oriented in space. Figure 6-5 illustrates the coordinate system for SensorTag sensors.

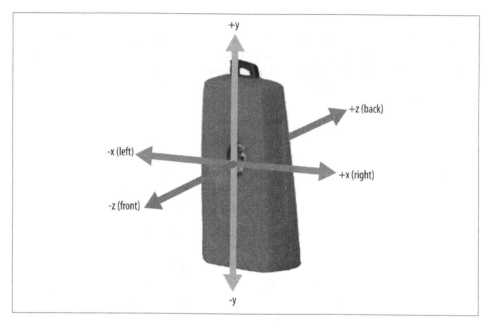

Figure 6-5. The coordinate system used by the SensorTag sensors

The Accelerometer

Now that we know how to connect to a Bluetooth low energy device, let's step back and take a closer look at the accelerometer we've been using up to this point.

What's an Accelerometer?

An accelerometer measures acceleration by detecting the force exerted on a mass, then uses this equation:

$$f = ma$$

to find the acceleration. The exact way the force and mass are measured varies a lot from one brand of sensor to another, from measuring the capacitance as the plates making up the capacitor are deformed to looking at the voltage generated by a crystal as it is

---

deformed. The SensorTag uses the KXTJ9 accelerometer from Kionox. It's a three-axis accelerometer with a selectable range of ±2G, ±4G, or ±8G. The SensorTag's firmware uses the ±2G setting, exposing 8 bits of the 14-bit resolution of the chip and giving a precision of 0.015625G.

Accessing the Accelerometer

Looking at the Generic Attribute Profile (GATT) for the accelerometer in Figure 6-6, we find three attributes in the accelerometer service. The GATT profile shown is a short excerpt from the one published by Texas Instruments. The four-digit values in the third column are the unique digits in the full 128-bit UUID for each service or characteristic. In all cases, the complete UUID is F000*xxxx*-0451-4000-B000-000000000000, where *xxxx* is replaced by the value from the GATT profile.

0x2B	43	0x2800	GATT_PRIMARY_SERVICE_UUID	**0xAA10 (ACCELEROMETER_SERV_UUID)**	GATT_PERMIT_READ	Start of Sensor Profile Accelerometer Service
0x2C		0x2803	GATT_CHARACTER_UUID	12 (properties: read/notify) 2D 00 (handle: 0x002D) 11 AA (UUID: 0xAA11)	GATT_PERMIT_READ	
0x2D	44 45	0xAA11	ACCELEROMETER_DATA_UUID	00:00:00 (3 bytes)	GATT_PERMIT_READ	X : Y : Z Coordinates
0x2E	46	0x2902	GATT_CLIENT_CHAR_CFG_UUID	00:00 (2 bytes)	GATT_PERMIT_READ / GATT_PERMIT_WRITE	Write "01:00" to enable notifications, "00:00" to disable
0x2F	47	0x2901	GATT_CHAR_USER_DESC_UUID	"Accel_Data" (14 bytes)	GATT_PERMIT_READ	
0x30	48	0x2803	GATT_CHARACTER_UUID	0A (properties: read/write) 31 00 (handle: 0x0031) 12 AA (UUID: 0xAA12)	GATT_PERMIT_READ	
0x31	49	0xAA12	ACCELEROMETER_CONF_UUID	1 (1 byte)	GATT_PERMIT_READ / GATT_PERMIT_WRITE	Write "01" to start Sensor and Measurements, "00" to put to sleep
0x32	50	0x2901	GATT_CHAR_USER_DESC_UUID	"Accel_Conf." (15 bytes)	GATT_PERMIT_READ	
0x33	51	0x2803	GATT_CHARACTER_UUID	0A (properties: read/write) 34 00 (handle: 0x0034) 13 AA (UUID: 0xAA13)	GATT_PERMIT_READ	
0x34	52	0xAA13	ACCELEROMETER_PERI_UUID	1 (1 byte)	GATT_PERMIT_READ / GATT_PERMIT_WRITE	Period = [input*10] ms, default 1000 ms, lower limit 100 ms
0x35	53	0x2901	GATT_CHAR_USER_DESC_UUID	"Acc_Period" (12 bytes)	GATT_PERMIT_READ	

Figure 6-6. Accelerometer GATT profile, courtesy of Texas Instruments

The complete program for accessing the accelerometer is listed at the end of this section for reference. It's also included in techBASIC and techBASIC Sampler. Look for the app called SensorTag Accelerometer in the *O'Reilly Books* folder.

The SensorTag program we looked at at the start of the chapter also reads the accelerometer, displaying it using an interactive graph in a nice GUI interface. It's in the *O'Reilly Books* folder as TI SensorTag.

Like all sensors, the accelerometer draws power. Keeping it off when it's not in use helps prolong the life of the battery. The AA12 attribute is used to turn the accelerometer on and off. Our program just turns it on:

```
CASE "F000AA12" & servicesHeader
  ! Turn the accelerometer sensor on.
  IF debug THEN PRINT "Accelerometer on."
  DIM value(1) as INTEGER
```

```
value(1) = 1
peripheral.writeCharacteristic(characteristics(i), value, 1)
```

Pass a 0 instead of a 1 to turn the accelerometer back off.

AA11 is used to read the accelerometer. There are two ways to read the acceleration, either reading it when the program wants an acceleration value or asking the accelerometer to notify the program whenever a new sensor reading is available. The first step is to tell the accelerometer which way we want to read values. To receive notifications when a value is available, write a 0 and a 1 to the characteristic, then ask it to start sending notifications:

```
SELECT CASE characteristics(i).uuid
  CASE "F000AA11" & servicesHeader
    ! Tell the accelerometer to begin sending data.
    IF debug THEN PRINT "Start accelerometer."
    DIM value(2) as INTEGER
    value = [0, 1]
    peripheral.writeCharacteristic(characteristics(i), value, 0)
    peripheral.setNotify(characteristics(i), 1)
```

The last parameter to writeCharacteristic tells the device if our program would like to get a response back after the value is received. For our case, it's easiest not to ask for a response, so the program passes 0. Some devices insist on sending back a response. In that case, a call will be made to BLECharacteristicInfo when the data has been received.

To read a single value, write two zero bytes to this characteristic:

```
DIM value(2) as INTEGER
value = [0, 0]
peripheral.writeCharacteristic(characteristics(i), value, 0)
```

Then, whenever a value is needed, read the characteristic:

```
peripheral.readCharacteristic(characteristics(i), 1)
```

The accelerometer sends samples about once a second unless you tell it otherwise. Use AA13 to set the sample rate. The value is expressed in tens of milliseconds, so passing 100 gives the default sample rate of once per second. Keep in mind that the hardware uses the sample rate as a suggestion, not a firm value! If the time between samples is really important, record the time with the sample value.

Whether you read the data once using readCharacteristic or ask for notifications with setNotify, the values are always returned using a call to BLECharacteristicInfo:

```
SUB BLECharacteristicInfo (time AS DOUBLE, _
                           peripheral AS BLEPeripheral, _
                           characteristic AS BLECharacteristic, _
                           kind AS INTEGER, _
                           message AS STRING, _
```

```
                          err AS LONG)
  IF kind = 2 THEN
    DIM value(1) AS INTEGER
    value = characteristic.value
    SELECT CASE characteristic.uuid
      CASE "F000AA11" & servicesHeader
        ! Update the accelerometer.
        c = 64.0
        p% = value(1)
        IF p% BITAND $0080 THEN p% = p% BITOR $FF00
        x = p%/c

        p% = value(2)
        IF p% BITAND $0080 THEN p% = p% BITOR $FF00
        y = p%/c

        p% = value(3)
        IF p% BITAND $0080 THEN p% = p% BITOR $FF00
        z = p%/c

        PRINT x, y, z
      CASE ELSE
        PRINT "Read from "; characteristic.uuid

    END SELECT
  ELSE IF kind = 3 AND err <> 0 THEN
    PRINT "Error writing "; characteristic.uuid; ": ("; err; ") "; message
  END IF
  END SUB
```

The value itself is returned as three bytes, where each byte is a signed value containing the acceleration along one axis. Naturally enough, the order is x, y, and z. Dividing by 64 converts from the range of the byte value, with a range of −128 to 127, to G force, with a range of about −2G to 2G.

Using the Accelerometer

The first step in using any sensor is to calibrate it. There are differences in individual sensors, and converting from a byte value to ±2G by dividing by 64 isn't perfect. After all, we get 128 values to the left of zero, and 127 on the positive side—it can't be exact.

Fortunately, we live on a pretty good calibration device. The surface of the Earth is, by definition, 1G. Lay the accelerometer on the table and run the program. The device may not be perfectly flat, but you can get the overall acceleration this way:

$$a = \sqrt{x^2 + y^2 + z^2}$$

or, in BASIC:

```
a = SQR(x*x + y*y + z*z)
```

If you like, collect a few hundred measurements and average the results. They should be 1G. Are they? If not, adjust all of your measurements by an appropriate factor so they are.

If you're really up on your physics, you may realize that the Earth's gravity is not the same strength everywhere. It varies because the Earth is not perfectly round, because it spins, and because of mineral deposits. This variation is smaller than 1%, though, and the precision of the accelerometer is about 1.5% of a gravity, so it's an error we can safely ignore.

One of the first things a lot of people try to do with an accelerometer is measure distance or speed. Let's see how that works by working through a simple example. Many people are familiar with one of the basic formulas of physics that allows us to calculate the distance something has traveled based on the time and acceleration. Here it is:

$$d = d_0 + v_0 t + \frac{1}{2} a t^2$$

To do anything useful, we also need to remember that:

$$v = v_0 + at$$

The accelerometer gives us acceleration. To keep things simple, let's assume we get back one sample every 0.1 seconds from sliding the SensorTag across a table, and the samples are 0.1G, 0.2G, and 0.3G along the direction we slide the SensorTag. Using this equation, and starting at rest, after one-tenth of a second:

$$v_1 = 0 + 0.1 \cdot 9.81 \cdot 0.1 = 0.0981 \; m/s$$
$$d_1 = 0 + 0t + \frac{1}{2} 0.1 \cdot 9.81 \cdot 0.1^2 = 0.004905 \; m$$

The factor of 9.81 converts from gravities to meters per second squared.

Continuing for the next two measurements:

$$v_2 = 0.0981 + 0.2 \cdot 9.81 \cdot 0.1 = 0.2943 \; m/s$$
$$d_2 = 0.004905 + 0.0981t + \frac{1}{2} 0.2 \cdot 9.81 \cdot 0.1^2 = 0.024525 \; m$$
$$v_3 = 0.2943 + 0.3 \cdot 9.81 \cdot 0.1 = 0.5886 \; m/s$$
$$d_3 = 0.024525 + 0.2943t + \frac{1}{2} 0.3 \cdot 9.81 \cdot 0.1^2 = 0.06867 \; m$$

After 0.3 seconds, the SensorTag has traveled about 7 cm (almost three inches) across the table, and it's traveling at a little over a half meter per second.

So now you're equipped to tape a SensorTag to any moving object and track where it is and how fast it's going, right? Ah, if only life were that simple. It works fine in theory, but in practice, an accelerometer is not a good way to track movement. The problem is the errors that crop up. There are two kinds: one is random error, where the measurements are off by a bit, but vary around the correct reading; the other is systematic error, where the values are off in a specific direction. Real measurements have both kinds of error, and they add up fast. The longer you take measurements, the more error will accumulate. While I personally find the whole subject of error analysis fascinating, that's not what this book is about—and you may not share my fascination. If you'd like to know more about these topics, start with web searches on systematic error, random error, and error analysis. The short version, though, is that calculating velocity and distance using an accelerometer is going to work a lot better for short, rapid movements like a rocket flight than for long, slow ones like tracking the progress of a robot. Error is still an issue for short, rapid movements, too; it's just not as big an issue.

The Source

Here's the source for a short accelerometer program. It's pretty simple, writing the values to the console. This is a great program to experiment a bit with the accelerometer. For another, see the source code for the SensorTag app, which displays all six SensorTag sensors in a nice GUI environment. This program is also included in both techBASIC and techBASIC Sampler. It's called SensorTag Accelerometer, and like the SensorTag app, it is in the *O'Reilly Books* folder.

```
! Simple program to access the accelerometer on the TI SensorTag.

! Set up variables to hold the peripheral and the characteristics
! for the battery and buzzer.
DIM sensorTag AS BLEPeripheral

! We will look for these services.
DIM servicesHeader AS STRING, services(1) AS STRING
servicesHeader = "-0451-4000-B000-000000000000"
services(1) = "F000AA10" & servicesHeader : ! Accelerometer
accel% = 1

! Start the BLE service and begin scanning for devices.
debug = 0
BLE.startBLE
DIM uuid(0) AS STRING
BLE.startScan(uuid)


! Called when a peripheral is found. If it is a Sensor Tag, we
! initiate a connection to it and stop scanning for peripherals.
!
! Parameters:
!    time - The time when the peripheral was discovered.
!    peripheral - The peripheral that was discovered.
!    services - List of services offered by the device.
!    advertisements - Advertisements (information provided by the
!        device without the need to read a service/characteristic)
!    rssi - Received Signal Strength Indicator
!
```

```
SUB BLEDiscoveredPeripheral (time AS DOUBLE, _
                              peripheral AS BLEPeripheral, _
                              services() AS STRING, _
                              advertisements(,) AS STRING, _
                              rssi)
IF peripheral.bleName = "TI BLE Sensor Tag" THEN
  sensorTag = peripheral
  BLE.connect(sensorTag)
  BLE.stopScan
END IF
END SUB


! Called to report information about the connection status of the
! peripheral or to report that services have been discovered.
!
! Parameters:
!    time - The time when the information was received.
!    peripheral - The peripheral.
!    kind - The kind of call. One of
!         1 - Connection completed
!         2 - Connection failed
!         3 - Connection lost
!         4 - Services discovered
!    message - For errors, a human-readable error message.
!    err - If there was an error, the Apple error number. If there
!         was no error, this value is 0.
!
SUB BLEPeripheralInfo (time AS DOUBLE, _
                        peripheral AS BLEPeripheral, _
                        kind AS INTEGER, _
                        message AS STRING, _
                        err AS LONG)
IF kind = 1 THEN
  ! The connection was established. Look for available services.
  IF debug THEN PRINT "Connection made."
  peripheral.discoverServices(uuid)
ELSE IF kind = 2 OR kind = 3 THEN
  IF debug THEN PRINT "Connection lost: "; kind
  BLE.connect(sensorTag)
ELSE IF kind = 4 THEN
  ! Services were found. If it is one of the ones we are interested
  ! in, begin discovery of its characteristics.
  DIM availableServices(1) AS BLEService
  availableServices = peripheral.services
  FOR s = 1 to UBOUND(services, 1)
    FOR a = 1 TO UBOUND(availableServices, 1)
      IF services(s) = availableServices(a).uuid THEN
        IF debug THEN PRINT "Discovering characteristics for "; services(s)
        peripheral.discoverCharacteristics(uuid, availableServices(a))
      END IF
    NEXT
  NEXT
```

```
END IF
END SUB


! Called to report information about a characteristic or included
! services for a service. If it is one we are interested in, start
! handling it.
!
! Parameters:
!    time - The time when the information was received.
!    peripheral - The peripheral.
!    service - The service whose characteristic or included
!        service was found.
!    kind - The kind of call. One of
!        1 - Characteristics found
!        2 - Included services found
!    message - For errors, a human-readable error message.
!    err - If there was an error, the Apple error number. If there
!        was no error, this value is 0.
!
SUB BLEServiceInfo (time AS DOUBLE, _
                    peripheral AS BLEPeripheral, _
                    service AS BLEService, _
                    kind AS INTEGER, _
                    message AS STRING, _
                    err AS LONG)
IF kind = 1 THEN
  ! Get the characteristics.
  DIM characteristics(1) AS BLECharacteristic
  characteristics = service.characteristics
  FOR i = 1 TO UBOUND(characteristics, 1)
    IF service.uuid = services(accel%) THEN
      ! Found the accelerometer.
      SELECT CASE characteristics(i).uuid
        CASE "F000AA11" & servicesHeader
          ! Tell the accelerometer to begin sending data.
          IF debug THEN PRINT "Start accelerometer."
          DIM value(2) as INTEGER
          value = [0, 1]
          peripheral.writeCharacteristic(characteristics(i), value, 0)
          peripheral.setNotify(characteristics(i), 1)

        CASE "F000AA12" & servicesHeader
          ! Turn the accelerometer sensor on.
          IF debug THEN PRINT "Accelerometer on."
          DIM value(1) as INTEGER
          value(1) = 1
          peripheral.writeCharacteristic(characteristics(i), value, 1)

        CASE "F000AA13" & servicesHeader
          ! Set the sample rate to 500ms.
          DIM value(1) as INTEGER
```

```
            value(1) = 50
            IF debug THEN PRINT "Setting accelerometer sample rate to "; value(1)
            peripheral.writeCharacteristic(characteristics(i), value, 1)
        END SELECT
      END IF
  NEXT
END IF
END SUB


! Called to return information from a characteristic.
!
! Parameters:
!   time - The time when the information was received.
!   peripheral - The peripheral.
!   characteristic - The characteristic whose information
!       changed.
!   kind - The kind of call. One of
!       1 - Called after a discoverDescriptors call.
!       2 - Called after a readCharacteristics call.
!       3 - Called to report status after a writeCharacteristics
!           call.
!   message - For errors, a human-readable error message.
!   err - If there was an error, the Apple error number. If there
!       was no error, this value is 0.
!
SUB BLECharacteristicInfo (time AS DOUBLE, _
                           peripheral AS BLEPeripheral, _
                           characteristic AS BLECharacteristic, _
                           kind AS INTEGER, _
                           message AS STRING, _
                           err AS LONG)
IF kind = 2 THEN
  DIM value(1) AS INTEGER
  value = characteristic.value
  SELECT CASE characteristic.uuid
    CASE "F000AA11" & servicesHeader
      ! Update the accelerometer.
      c = 64.0
      p% = value(1)
      IF p% BITAND $0080 THEN p% = p% BITOR $FF00
      x = p%/c

      p% = value(2)
      IF p% BITAND $0080 THEN p% = p% BITOR $FF00
      y = p%/c

      p% = value(3)
      IF p% BITAND $0080 THEN p% = p% BITOR $FF00
      z = p%/c

      PRINT x, y, z
```

```
      CASE ELSE
          PRINT "Read from "; characteristic.uuid

      END SELECT
    ELSE IF kind = 3 AND err <> 0 THEN
      PRINT "Error writing "; characteristic.uuid; ": ("; err; ") "; message
    END IF
END SUB
```

The Barometer

A barometer measures the pressure exerted by a gas or liquid. The T5400 Digital Barometric Pressure Sensor used on the SensorTag is designed for use in air at pressures
from 0.3 atmospheres to 1.1 atmospheres. There are a lot of competing units for pressure, including pascals or hectopascals (hundreds of pascals), pounds per square inch,
Torr, millimeters of mercury, and bars. One bar is roughly one standard atmosphere,
and since our barometer is designed for atmospheric pressure, that's the unit we'll use.

There are a lot of reasons for the pressure in the atmosphere to vary. One is altitude.
Barometers are a common way to measure the altitude of anything that flies. We're a
mile high here in Albuquerque, where the barometric pressure is 0.83 bar today. Another
common reason for air pressure to vary is weather. A pressure change is common as a
weather front moves through. The pressure drops as a wet weather system approaches,
then climbs again after it passes by.

The barometer needs to know the temperature to calculate the pressure, so the T5400
is actually two sensors: it can return the temperature as well as pressure.

Accessing the Barometer

Figure 6-7 shows the Generic Attribute Profile (GATT) for the barometer service.

0x49	73	0x2800	GATT_PRIMARY_SERVICE_UUID	0xAA40 (BAROMETER_SERV_UUID)		GATT_PERMIT_READ	Start of Sensor Profile Barometer Service
0x4A		0x2803	GATT_CHARACTER_UUID	12 (properties: read/notify) 4B 00 (handle: 0x004B) 41 AA (UUID: 0xAA41)		GATT_PERMIT_READ	
0x4B	74 75	0xAA41	BAROMETER_DATA_UUID	00:00:00:00 (4 bytes)		GATT_PERMIT_READ	TempLSB:TempMSB:PressureLSB:PressureMSB
0x4C	76	0x2902	GATT_CLIENT_CHAR_CFG_UUID	00:00 (2 bytes)		GATT_PERMIT_READ GATT_PERMIT_WRITE	
0x4D	77	0x2901	GATT_CHAR_USER_DESC_UUID	"Barometer Data." (15 bytes)		GATT_PERMIT_READ	
0x4E	78	0x2803	GATT_CHARACTER_UUID	0A (properties: read/write) 53 00 (handle: 0x0053) 42 AA (UUID: 0xAA42)		GATT_PERMIT_READ	
0x4F	79	0xAA42	BAROMETER_CONF_UUID	1 (1 byte)		GATT_PERMIT_READ GATT_PERMIT_WRITE	Write "01" to start Sensor and Measurements, "00" to put to sleep, "02" to read calibration values from sensor
0x50	80	0x2901	GATT_CHAR_USER_DESC_UUID	"Barometer Conf." (16 bytes)		GATT_PERMIT_READ	
0x51	81	0x2803	GATT_CHARACTER_UUID	02 (properties: read only) 4F 00 (handle: 0x004F) 43 AA (UUID: 0xAA43)		GATT_PERMIT_READ	
0x52	82	0xAA43	BAROMETER_CALI_UUID	00:00:...:00:00 (16 bytes)		GATT_PERMIT_READ	When write 02 to Barometer Conf. has been issued, the calibration values is found here
0x53	83	0x2902	GATT_CLIENT_CHAR_CFG_UUID	00:00 (2 bytes)		GATT_PERMIT_READ GATT_PERMIT_WRITE	
0x54	84	0x2901	GATT_CHAR_USER_DESC_UUID	"Barometer Cali." (16 bytes)		GATT_PERMIT_READ	

Figure 6-7. Barometer GATT profile, courtesy of Texas Instruments

The complete program for accessing the barometer is listed at the end of this section for reference. It's also included in techBASIC and techBASIC Sampler. Look for the app called SensorTag Barometer in the *O'Reilly Books* folder.

The SensorTag program from the start of the chapter also reads the barometer, displaying it using a nice GUI interface. It's in the *O'Reilly Books* folder as TI SensorTag.

As with the other sensors on the SensorTag, the second attribute, AA42, turns the sensor on or off. Unlike with the other sensors, however, it has another use. Several values are needed to calculate the barometric pressure from the raw data returned by the sensor. Writing a 2 to attribute AA42 tells the barometer to calculate these calibration values. Here's the code to turn the sensor on and ask it to find the calibration values:

```
CASE "F000AA42" & servicesHeader
  ! Turn the pressure sensor on.
  IF debug THEN PRINT "Pressure on."
  DIM value(1) as INTEGER
  value(1) = 1
  peripheral.writeCharacteristic(characteristics(i), value, 1)
  value(1) = 2
  peripheral.writeCharacteristic(characteristics(i), value, 1)
```

You can turn the barometer off by writing a 0.

Calibration data is needed before any of the sensor data can be used. The AA43 attribute is used to read the calibration data. Here's the code to ask the device to send back the calibration data:

```
CASE "F000AA43" & servicesHeader
  ! Get the calibration data.
  peripheral.readCharacteristic(characteristics(i))
```

The calibration data returned is an array of 16 bytes making up eight of the values. All values are two-byte integers stored least significant byte first, but the first four are unsigned values and the last four are signed. Here's the code from BLECharacteristicInfo that reads the calibration data and stores it for later use:

```
CASE "F000AA43" & servicesHeader
  ! Get the pressure calibration data.
  IF debug THEN PRINT "Calibration data read."
  FOR i = 1 TO 4
    j = 1 + (i - 1)*2
        m_barCalib(i) = (value(j) BITOR (value(j + 1) << 8)) BITAND $00FFFF
  NEXT
  FOR i = 5 TO 8
        j = 1 + (i - 1)*2
```

```
    m_barCalib(i) = value(j) BITOR (value(j + 1) << 8)
NEXT
```

More Bit and Byte Manipulation

The SensorTag sends back a two-byte integer. In one case, it is a signed value that is the same length as a techBASIC two-byte integer, so the conversion from the two bytes it sends back to the techBASIC value is fairly straightforward.

The first byte it sends is the least significant eight bits of the value. Putting this in more familiar terms, if the value was sent back as a decimal number and the number was 67, the SensorTag would send back two digits, a 7 followed by a 6. The 7 is the least significant digit, just as the first byte is the least significant byte. The program uses this byte just as it comes back when it grabs value(j).

The second byte is the most significant byte. Returning to the decimal example, it's a 6, but it represents 60 in the full number. The program needs to multiply the most significant byte by 256, since we're using bytes rather than decimal digits. That's what value(j + 1) << 8 does—by shifting the value left by 8 bits, it multiplies the byte by 256.

Finally, the two pieces of the number must be put together. That's what the BITOR operation does.

That works fine for a signed two-byte value that, like the techBASIC integer, represents numbers from –32,768 to 32,767. An unsigned two-byte value can range from 0 to 65,535, though, and that's too big for a techBASIC integer. The program converts the value to a long integer, instead. techBASIC long integers can range from –2,147,483,648 to 2,147,483,647—more than enough to handle an unsigned number that tops out at 65,535. The program handles this by converting the number to a two-byte value the same way it does for a signed number, followed by using the BITAND $00FFFF operation. This does two things. First, since $00FFFF is more than four hexadecimal digits, techBASIC treats it as a long integer, and converts the two-byte value to a long integer, too. Second, by only accepting the least significant two bytes of the value, the signed value is converted back to the proper positive number.

There are two ways to read the barometer, either reading it when the program wants the pressure or asking the barometer to notify the program whenever a new sensor reading is available. The next step is to tell the barometer which way we want to read values. Write a 1 and a 0 to the characteristic to receive notifications when a value is available, then ask it to start sending notifications:

```
CASE "F000AA41" & servicesHeader
  ! Tell the pressure sensor to begin sending data.
  IF debug THEN PRINT "Start pressure sensor."
  DIM value(2) as INTEGER
  value = [0, 1]
```

```
peripheral.writeCharacteristic(characteristics(i), value, 0)
peripheral.setNotify(characteristics(i), 1)
```

To read a single value, write two zero bytes to this characteristic:

```
DIM value(2) as INTEGER
value = [0, 0]
peripheral.writeCharacteristic(characteristics(i), value, 0)
```

Then, whenever a value is needed, read the characteristic:

```
peripheral.readCharacteristic(characteristics(i), 1)
```

The barometer can take several seconds to return the first value, so be patient. This program also doesn't check to make sure the calibration data has been read before processing the barometer readings, so the first value reported may be zero. The values will become reasonable once the calibration data has been read.

Whether you read the data once using readCharacteristic or ask for notifications with setNotify, the values are always returned using a call to BLECharacteristicInfo. The values returned consist of a temperature and the raw pressure reading, both stored in two-byte integers with the least significant byte first. The temperature is signed, while the pressure is an unsigned value. Here's the complete set of equations to convert the raw temperature and pressure to an atmospheric pressure using the eight calibration values:

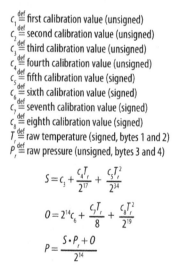

$$c_1 \overset{\text{def}}{=} \text{first calibration value (unsigned)}$$
$$c_2 \overset{\text{def}}{=} \text{second calibration value (unsigned)}$$
$$c_3 \overset{\text{def}}{=} \text{third calibration value (unsigned)}$$
$$c_4 \overset{\text{def}}{=} \text{fourth calibration value (unsigned)}$$
$$c_5 \overset{\text{def}}{=} \text{fifth calibration value (signed)}$$
$$c_6 \overset{\text{def}}{=} \text{sixth calibration value (signed)}$$
$$c_7 \overset{\text{def}}{=} \text{seventh calibration value (signed)}$$
$$c_8 \overset{\text{def}}{=} \text{eighth calibration value (signed)}$$
$$T_r \overset{\text{def}}{=} \text{raw temperature (signed, bytes 1 and 2)}$$
$$P_r \overset{\text{def}}{=} \text{raw pressure (unsigned, bytes 3 and 4)}$$

$$S = c_3 + \frac{c_4 T_r}{2^{17}} + \frac{c_5 T_r^2}{2^{34}}$$

$$O = 2^{14} c_6 + \frac{c_7 T_r}{8} + \frac{c_8 T_r^2}{2^{19}}$$

$$P = \frac{S \cdot P_r + O}{2^{14}}$$

P is the pressure in pascals; divide by 100,000 to convert from pascals to bars.

You might have noticed that there are eight calibration constants, but we only used six. The first two are used to convert the raw temperature to degrees Celsius using the formula:

$$T = \frac{c_2}{2^{10}} + \frac{c_1 T_r}{2^{24}}$$

Here's what we get after converting the equations to BASIC:

```
SUB BLECharacteristicInfo (time AS DOUBLE, _
                           peripheral AS BLEPeripheral, _
                           characteristic AS BLECharacteristic, _
                           kind AS INTEGER, _
                           message AS STRING, _
                           err AS LONG)
IF kind = 2 THEN
  DIM value(1) AS INTEGER
  value = characteristic.value
  SELECT CASE characteristic.uuid
    CASE "F000AA41" & servicesHeader
      ! Update the pressure indicator.
      Tr = value(1) BITOR (value(2) << 8)
      S = m_barCalib(3) + Tr*(m_barCalib(4)/2^17 + Tr*m_barCalib(5)/2^34)
      O = m_barCalib(6)*2^14 + Tr*(m_barCalib(7)/8.0 + Tr*m_barCalib(8)/2^19)
      Pr = (value(3) BITOR (value(4) << 8)) BITAND $00FFFF
      Pa = (S*Pr + O)/2^14

      ! Convert from Pascal to Bar.
      Pa = Pa/100000.0
      PRINT "Pressure: "; Pa; " Bar";

      ! Add the temperature.
      T = m_barCalib(2)/2^10 + Tr*m_barCalib(1)/2^24
      PRINT ", temperature: "; T

    CASE "F000AA43" & servicesHeader
      ! Get the pressure calibration data.
      IF debug THEN PRINT "Calibration data read."
      FOR i = 1 TO 4
        j = 1 + (i - 1)*2
        m_barCalib(i) = (value(j) BITOR (value(j + 1) << 8)) BITAND $00FFFF
      NEXT
      FOR i = 5 TO 8
        j = 1 + (i - 1)*2
        m_barCalib(i) = value(j) BITOR (value(j + 1) << 8)
      NEXT

    CASE ELSE
      PRINT "Read from "; characteristic.uuid

  END SELECT
ELSE IF kind = 3 AND err <> 0 THEN
  PRINT "Error writing "; characteristic.uuid; ": ("; err; ") "; message
END IF
END SUB
```

Running the program gave the following for the first few measurements. Note that, as expected, the first value is not usable, since the barometer had not quite warmed up. The temperature is in Celsius, so it really wasn't all that cold:

```
Pressure: 0 Bar, temperature: 0
Pressure: 0.837669 Bar, temperature: 22.870993
Pressure: 0.831656 Bar, temperature: 22.870993
```

The Source

Here's the complete source for the barometer program:

```
! Simple program to access the barometer on the TI SensorTag.

! Set up variables to hold the peripheral and the characteristics
! for the battery and buzzer.
DIM sensorTag AS BLEPeripheral

! We will look for these services.
DIM servicesHeader AS STRING, services(1) AS STRING
servicesHeader = "-0451-4000-B000-000000000000"
services(1) = "F000AA40" & servicesHeader : ! Pressure
press% = 1

! Start the BLE service and begin scanning for devices.
debug = 0
BLE.startBLE
DIM uuid(0) AS STRING
BLE.startScan(uuid)

! Create a place for the barometer calibration values.
DIM m_barCalib(8)

! Called when a peripheral is found. If it is a Sensor Tag, we
! initiate a connection to it and stop scanning for peripherals.
!
! Parameters:
!    time - The time when the peripheral was discovered.
!    peripheral - The peripheral that was discovered.
!    services - List of services offered by the device.
!    advertisements - Advertisements (information provided by the
!        device without the need to read a service/characteristic)
!    rssi - Received Signal Strength Indicator
!
SUB BLEDiscoveredPeripheral (time AS DOUBLE, _
                             peripheral AS BLEPeripheral, _
                             services() AS STRING, _
                             advertisements(,) AS STRING, _
                             rssi)
IF peripheral.bleName = "TI BLE Sensor Tag" THEN
   sensorTag = peripheral
   BLE.connect(sensorTag)
```

```
      BLE.stopScan
END IF
END SUB

! Called to report information about the connection status of the
! peripheral or to report that services have been discovered.
!
! Parameters:
!     time - The time when the information was received.
!     peripheral - The peripheral.
!     kind - The kind of call. One of
!          1 - Connection completed
!          2 - Connection failed
!          3 - Connection lost
!          4 - Services discovered
!     message - For errors, a human-readable error message.
!     err - If there was an error, the Apple error number. If there
!          was no error, this value is 0.
!
SUB BLEPeripheralInfo (time AS DOUBLE, _
                       peripheral AS BLEPeripheral, _
                       kind AS INTEGER, _
                       message AS STRING, _
                       err AS LONG)
IF kind = 1 THEN
   ! The connection was established. Look for available services.
   IF debug THEN PRINT "Connection made."
   peripheral.discoverServices(uuid)
ELSE IF kind = 2 OR kind = 3 THEN
   IF debug THEN PRINT "Connection lost: "; kind
   BLE.connect(sensorTag)
ELSE IF kind = 4 THEN
   ! Services were found. If it is one of the ones we are interested
   ! in, begin discovery of its characteristics.
   DIM availableServices(1) AS BLEService
   availableServices = peripheral.services
   FOR s = 1 to UBOUND(services, 1)
     FOR a = 1 TO UBOUND(availableServices, 1)
       IF services(s) = availableServices(a).uuid THEN
         IF debug THEN PRINT "Discovering characteristics for "; services(s)
         peripheral.discoverCharacteristics(uuid, availableServices(a))
       END IF
     NEXT
   NEXT
END IF
END SUB

! Called to report information about a characteristic or included
! services for a service. If it is one we are interested in, start
! handling it.
!
```

```
! Parameters:
!    time - The time when the information was received.
!    peripheral - The peripheral.
!    service - The service whose characteristic or included
!        service was found.
!    kind - The kind of call. One of
!        1 - Characteristics found
!        2 - Included services found
!    message - For errors, a human-readable error message.
!    err - If there was an error, the Apple error number. If there
!        was no error, this value is 0.
!
SUB BLEServiceInfo (time AS DOUBLE, _
                    peripheral AS BLEPeripheral, _
                    service AS BLEService, _
                    kind AS INTEGER, _
                    message AS STRING, _
                    err AS LONG)
IF kind = 1 THEN
  ! Get the characteristics.
  DIM characteristics(1) AS BLECharacteristic
  characteristics = service.characteristics
  FOR i = 1 TO UBOUND(characteristics, 1)
    IF service.uuid = services(press%) THEN
      ! Found the pressure sensor.
      SELECT CASE characteristics(i).uuid
        CASE "F000AA41" & servicesHeader
          ! Tell the pressure sensor to begin sending data.
          IF debug THEN PRINT "Start pressure sensor."
          DIM value(2) as INTEGER
          value = [0, 1]
          peripheral.writeCharacteristic(characteristics(i), value, 0)
          peripheral.setNotify(characteristics(i), 1)

        CASE "F000AA42" & servicesHeader
          ! Turn the pressure sensor on.
          IF debug THEN PRINT "Pressure on."
          DIM value(1) as INTEGER
          value(1) = 1
          peripheral.writeCharacteristic(characteristics(i), value, 1)
          value(1) = 2
          peripheral.writeCharacteristic(characteristics(i), value, 1)

        CASE "F000AA43" & servicesHeader
          ! Get the calibration data.
          peripheral.readCharacteristic(characteristics(i))
      END SELECT
    END IF
  NEXT
END IF
END SUB
```

```
! Called to return information from a characteristic.
!
! Parameters:
!    time - The time when the information was received.
!    peripheral - The peripheral.
!    characteristic - The characteristic whose information
!        changed.
!    kind - The kind of call. One of
!        1 - Called after a discoverDescriptors call.
!        2 - Called after a readCharacteristics call.
!        3 - Called to report status after a writeCharacteristics
!            call.
!    message - For errors, a human-readable error message.
!    err - If there was an error, the Apple error number. If there
!        was no error, this value is 0.
!
SUB BLECharacteristicInfo (time AS DOUBLE, _
                          peripheral AS BLEPeripheral, _
                          characteristic AS BLECharacteristic, _
                          kind AS INTEGER, _
                          message AS STRING, _
                          err AS LONG)
IF kind = 2 THEN
  DIM value(1) AS INTEGER
  value = characteristic.value
  SELECT CASE characteristic.uuid
    CASE "F000AA41" & servicesHeader
      ! Update the pressure indicator.
      Tr = value(1) BITOR (value(2) << 8)
      S = m_barCalib(3) + Tr*(m_barCalib(4)/2^17 + Tr*m_barCalib(5)/2^34)
      O = m_barCalib(6)*2^14 + Tr*(m_barCalib(7)/8.0 + Tr*m_barCalib(8)/2^19)
      Pr = (value(3) BITOR (value(4) << 8)) BITAND $00FFFF
      Pa = (S*Pr + O)/2^14

      ! Convert from Pascal to Bar.
      Pa = Pa/100000.0
      PRINT "Pressure: "; Pa; " Bar";

      ! Add the temperature.
      T = m_barCalib(2)/2^10 + Tr*m_barCalib(1)/2^24
      PRINT ", temperature: "; T

    CASE "F000AA43" & servicesHeader
      ! Get the pressure calibration data.
      IF debug THEN PRINT "Calibration data read."
      FOR i = 1 TO 4
        j = 1 + (i - 1)*2
        m_barCalib(i) = (value(j) BITOR (value(j + 1) << 8)) BITAND $00FFFF
      NEXT
      FOR i = 5 TO 8
        j = 1 + (i - 1)*2
```

```
        m_barCalib(i) = value(j) BITOR (value(j + 1) << 8)
      NEXT

    CASE ELSE
      PRINT "Read from "; characteristic.uuid

  END SELECT
ELSE IF kind = 3 AND err <> 0 THEN
  PRINT "Error writing "; characteristic.uuid; ": ("; err; ") "; message
END IF
END SUB
```

The Gyroscope

A gyroscope measures how fast something rotates. The IMU-3000 Triple Axis Motion-Processor™ gyroscope used on the TI SensorTag returns rotation about three axes in degrees per second. The chip can return rotation rates across several ranges; the SensorTag preselects a range of ±250 degrees/second, giving a maximum precision in exchange for a smaller maximum rotation rate. The precision is about 0.007 degrees per second, although based on observations of a SensorTag that is stationary on my desk, I don't think it's really that accurate. Or perhaps my world really is spinning around randomly—it certainly feels that way on occasion. Based on casual observation, I'd say the true accuracy is about 0.2 degrees per second, and there is a definite need for calibration.

Accessing the Gyroscope

Looking at the Generic Attribute Profile (GATT) in Figure 6-8, you'll see two characteristics for the gyroscope.

0x55	85	0x2800	GATT_PRIMARY_SERVICE_UUID	0xAA50 (GYROSCOPE_SERV_UUID)	GATT_PERMIT_READ	Start of Sensor Profile Gyroscope Service
0x56	86	0x2803	GATT_CHARACTER_UUID	12 (properties: read/notify) 57 00 (handle: 0x0057) 51 AA (UUID: 0xAA51)	GATT_PERMIT_READ	
0x57	87	0xAA51	GYROSCOPE_DATA_UUID	00:00:00:00:00:00 (6 bytes)	GATT_PERMIT_READ	XMSB:XLSB:YMSB:YLSB: ZMSB:ZLSB
0x58	88	0x2902	GATT_CLIENT_CHAR_CFG_UUID	00:00 (2 bytes)	GATT_PERMIT_READ | GATT_PERMIT_WRITE	
0x59	89	0x2901	GATT_CHAR_USER_DESC_UUID	"Gyro. Data" (11 bytes)	GATT_PERMIT_READ	
0x5A	90	0x2803	GATT_CHARACTER_UUID	0A (properties: read/write) 5B 00 (handle: 0x005B) 52 AA (UUID: 0xAA52)	GATT_PERMIT_READ	
0x5B	91	0xAA52	GYROSCOPE_CONF_UUID	1 (1 byte)	GATT_PERMIT_READ | GATT_PERMIT_WRITE	Write 0 to turn off gyroscope, 1 to enable X axis only, 2 to enable Y axis only, 3 = X and Y, 4 = Z only, 5 = X and Z, 6 = Y and Z, 7 = X, Y and Z
0x5C	92	0x2901	GATT_CHAR_USER_DESC_UUID	"Gyro. Conf." (13 bytes)	GATT_PERMIT_READ	

Figure 6-8. Gyroscope GATT profile, courtesy of Texas Instruments

The complete program for accessing the gyroscope is listed at the end of this section for reference. It's also included in techBASIC and tech-BASIC Sampler. Look for the app called SensorTag Gyroscope in the *O'Reilly Books* folder.

The SensorTag program from the start of the chapter also reads the gyroscope, displaying it using an interactive graph in a nice GUI interface. It's in the *O'Reilly Books* folder as TI SensorTag.

The 0xAA52 characteristic is used to turn the gyroscope on and off. Our program just turns it on:

```
CASE "F000AA52" & servicesHeader
  ! Turn the gyroscope on.
  IF debug THEN PRINT "Gyroscope on."
  DIM value(1) as INTEGER
  value(1) = 7
  peripheral.writeCharacteristic(characteristics(i), value, 1)
```

The x-, y-, and z-axes can be turned on and off independently. Passing 7 turns them all on. Pass a 0 instead of a 7 to turn the gyroscope off.

More Bits and Bytes

Why does passing 7 turn the gyroscope on for all three axes, and how would you turn on a specific axis? That's all tied up in a common method used here called *bitmapping*.

Normally, a byte is used to represent the numbers from 0 to 255. Here, the bits are used for something else entirely. A byte is eight bits; let's represent them as eight 0 or 1 values, like this for a zero byte:

```
0000 0000
```

The least significant bit—the one on the right—is used to control the gyroscope's x-axis. Setting it to 1 would turn on the x-axis. The least significant bit also represents 1 in binary numbers, so:

```
0000 0001
```

is a value of 1 and controls the x-axis.

The second bit controls the y-axis. It has a value of 2 in a binary number, so:

```
0000 0010
```

turns the y-axis of the gyroscope on. techBASIC represents this as the number 2. The next bit, which has a value of 4 in a binary number, controls the z-axis:

```
0000 0100
```

Adding all of the values together in binary gives the bit map:

```
0000 0111
```

which will turn all three axes on or off. Adding the three values together in decimal gives 7, the value passed in the program.

AA51 is used to read the gyroscope. There are two ways to read the gyroscope, either reading it when the program wants a rotation rate or asking the gyroscope to notify the program whenever a new sensor reading is available. The first step is to tell the gyroscope which way we want to read values. To receive notifications when a value is available, write a 1 and a 0 to the characteristic, and then ask it to start sending notifications:

```
CASE "F000AA51" & servicesHeader
  ! Tell the gyroscope to begin sending data.
  IF debug THEN PRINT "Start gyroscope."
  DIM value(2) as INTEGER
  value = [0, 1]
  peripheral.writeCharacteristic(characteristics(i), value, 0)
  peripheral.setNotify(characteristics(i), 1)
```

To read a single value, write two zero bytes to this characteristic:

```
DIM value(2) as INTEGER
value = [0, 0]
peripheral.writeCharacteristic(characteristics(i), value, 0)
```

Then, whenever a value is needed, read the characteristic:

```
peripheral.readCharacteristic(characteristics(i), 1)
```

Whether you read the data once using readCharacteristic or ask for notifications with setNotify, the values are always returned using a call to BLECharacteristicInfo. The value reported consists of three two-byte signed integers, one each for the rate of rotation about the x-, y-, and z-axes, in that order. Each value is least significant byte first. The integers must be divided by 65,536 and multiplied by 500 to yield the final range of –250 to 250 degrees per second. Here's the BASIC code to take care of reading and printing the values:

```
SUB BLECharacteristicInfo (time AS DOUBLE, _
                           peripheral AS BLEPeripheral, _
                           characteristic AS BLECharacteristic, _
                           kind AS INTEGER, _
                           message AS STRING, _
                           err AS LONG)
  IF kind = 2 THEN
    DIM value(1) AS INTEGER
    value = characteristic.value
    SELECT CASE characteristic.uuid
      CASE "F000AA51" & servicesHeader
        ! Update the gyroscope.
```

```
      c = 65536.0/500.0
      x = ((value(2) << 8) BITOR value(1))/c
      y = ((value(4) << 8) BITOR value(3))/c
      z = ((value(6) << 8) BITOR value(5))/c
      PRINT time, x, y, z

    CASE ELSE
      PRINT "Read from "; characteristic.uuid

  END SELECT
  ELSE IF kind = 3 AND err <> 0 THEN
    PRINT "Error writing "; characteristic.uuid; ": ("; err; ") "; message
  END IF
END SUB
```

Using the Gyroscope

Calibrating the gyroscope proceeds in two stages. The first is to set the zero point. That's actually rather easy. Lay the SensorTag on the table and record a few dozen to a few hundred values. The average should be zero, since the device is not rotating. Subtract the actual average from each reading to force the mean to zero. Take the standard deviation of the samples if you need a good handle on the accuracy of the device.

Of course, that doesn't mean that a rotation rate of 100 degrees per second is really rotating at 100 degrees per second. If you would like to calibrate the scale, and not just the zero point, place the device on a slow motor from a toy and start it spinning. A motor that turns about twice a second should work well. You can count the rotations over a minute and use the value to calculate the actual rotation, comparing that to the reported rotation. If you really need this kind of calibration, test the device separately for each axis, and spin it in both directions. You might even want to spin it at several speeds.

Gyroscopes are used for all sorts of applications, from physics experiments to flying aircraft. They work really well when coupled with an accelerometer and magnetometer. The combination of inputs from all three sensors helps correct errors from any one of them alone. For example, a gyroscope might have a systematic error that, over time, would indicate it is turning slowly. Couple it with a compass, though, and it's easy to see there is no drift. At the same time, the compass is not terribly accurate for short, quick movements that occur in augmented reality applications or flight. Couple it with the gyroscope, and the rapid changes from a heading can be handled smoothly. The amount of work you put into using the sensors together really depends a lot on the application.

The Source

Now let's look at the source for the gyroscope program. It prints the time when the sample was returned and the rotation rate along each axis to the console, updating the values about once a second. This is a great program to experiment a bit with the gyro-scope. Here are the first few values printed when I ran the program:

```
385593255.313462972641    98.472588    -98.861687    171.607956
385593256.313488006592    -75.637817   -8.68988      -50.361629
385593257.313521027565    0.480652     14.755248     -34.355164
385593258.313567996025    -3.036499    -0.450134     -19.973755
385593259.313566029072    49.400326    71.83075       49.72076
385593260.31361502409     8.216858     12.535094      81.375114
385593261.313639998436    -11.428832   23.269651      46.913143
385593262.313674986362    17.303467    -5.119323      44.982906
```

And here's the source:

```
! Simple program to access the gyroscope on the TI SensorTag.

! Set up variables to hold the peripheral and the characteristics
! for the battery and buzzer.
DIM sensorTag AS BLEPeripheral

! We will look for these services.
DIM servicesHeader AS STRING, services(1) AS STRING
servicesHeader = "-0451-4000-B000-000000000000"
services(1) = "F000AA50" & servicesHeader : ! Gyroscope
gyro% = 1

! Start the BLE service and begin scanning for devices.
debug = 0
BLE.startBLE
DIM uuid(0) AS STRING
BLE.startScan(uuid)

! Called when a peripheral is found. If it is a Sensor Tag, we
! initiate a connection to it and stop scanning for peripherals.
!
! Parameters:
!    time - The time when the peripheral was discovered.
!    peripheral - The peripheral that was discovered.
!    services - List of services offered by the device.
!    advertisements - Advertisements (information provided by the
!        device without the need to read a service/characteristic)
!    rssi - Received Signal Strength Indicator
!
SUB BLEDiscoveredPeripheral (time AS DOUBLE, _
                            peripheral AS BLEPeripheral, _
                            services() AS STRING, _
                            advertisements(,) AS STRING, _
                            rssi)
IF peripheral.bleName = "TI BLE Sensor Tag" THEN
  sensorTag = peripheral
  BLE.connect(sensorTag)
  BLE.stopScan
END IF
```

```
END SUB

! Called to report information about the connection status of the
! peripheral or to report that services have been discovered.
!
! Parameters:
!    time - The time when the information was received.
!    peripheral - The peripheral.
!    kind - The kind of call. One of
!        1 - Connection completed
!        2 - Connection failed
!        3 - Connection lost
!        4 - Services discovered
!    message - For errors, a human-readable error message.
!    err - If there was an error, the Apple error number. If there
!        was no error, this value is 0.
!
SUB BLEPeripheralInfo (time AS DOUBLE, _
                       peripheral AS BLEPeripheral, _
                       kind AS INTEGER, _
                       message AS STRING, _
                       err AS LONG)
IF kind = 1 THEN
  ! The connection was established. Look for available services.
  IF debug THEN PRINT "Connection made."
  peripheral.discoverServices(uuid)
ELSE IF kind = 2 OR kind = 3 THEN
  IF debug THEN PRINT "Connection lost: "; kind
  BLE.connect(sensorTag)
ELSE IF kind = 4 THEN
  ! Services were found. If it is one of the ones we are interested
  ! in, begin discovery of its characteristics.
  DIM availableServices(1) AS BLEService
  availableServices = peripheral.services
  FOR s = 1 to UBOUND(services, 1)
    FOR a = 1 TO UBOUND(availableServices, 1)
      IF services(s) = availableServices(a).uuid THEN
        IF debug THEN PRINT "Discovering characteristics for "; services(s)
        peripheral.discoverCharacteristics(uuid, availableServices(a))
      END IF
    NEXT
  NEXT
END IF
END SUB


! Called to report information about a characteristic or included
! services for a service. If it is one we are interested in, start
! handling it.
!
! Parameters:
!    time - The time when the information was received.
```

```
!    peripheral - The peripheral.
!    service - The service whose characteristic or included
!        service was found.
!    kind - The kind of call. One of
!        1 - Characteristics found
!        2 - Included services found
!    message - For errors, a human-readable error message.
!    err - If there was an error, the Apple error number. If there
!        was no error, this value is 0.
!
SUB BLEServiceInfo (time AS DOUBLE, _
                    peripheral AS BLEPeripheral, _
                    service AS BLEService, _
                    kind AS INTEGER, _
                    message AS STRING, _
                    err AS LONG)
IF kind = 1 THEN
  ! Get the characteristics.
  DIM characteristics(1) AS BLECharacteristic
  characteristics = service.characteristics
  FOR i = 1 TO UBOUND(characteristics, 1)
    IF service.uuid = services(gyro%) THEN
      ! Found the gyroscope.
      SELECT CASE characteristics(i).uuid
        CASE "F000AA51" & servicesHeader
          ! Tell the gyroscope to begin sending data.
          IF debug THEN PRINT "Start gyroscope."
          DIM value(2) as INTEGER
          value = [0, 1]
          peripheral.writeCharacteristic(characteristics(i), value, 0)
          peripheral.setNotify(characteristics(i), 1)

        CASE "F000AA52" & servicesHeader
          ! Turn the gyroscope on.
          IF debug THEN PRINT "Gyroscope on."
          DIM value(1) as INTEGER
          value(1) = 7
          peripheral.writeCharacteristic(characteristics(i), value, 1)
      END SELECT
    END IF
  NEXT
END IF
END SUB


! Called to return information from a characteristic.
!
! Parameters:
!    time - The time when the information was received.
!    peripheral - The peripheral.
!    characteristic - The characteristic whose information
!        changed.
```

```
!     kind - The kind of call. One of
!          1 - Called after a discoverDescriptors call.
!          2 - Called after a readCharacteristics call.
!          3 - Called to report status after a writeCharacteristics
!              call.
!     message - For errors, a human-readable error message.
!     err - If there was an error, the Apple error number. If there
!          was no error, this value is 0.
!
SUB BLECharacteristicInfo (time AS DOUBLE, _
                           peripheral AS BLEPeripheral, _
                           characteristic AS BLECharacteristic, _
                           kind AS INTEGER, _
                           message AS STRING, _
                           err AS LONG)
IF kind = 2 THEN
  DIM value(1) AS INTEGER
  value = characteristic.value
  SELECT CASE characteristic.uuid
    CASE "F000AA51" & servicesHeader
      ! Update the gyroscope.
      c = 65536.0/500.0
      x = ((value(2) << 8) BITOR value(1))/c
      y = ((value(4) << 8) BITOR value(3))/c
      z = ((value(6) << 8) BITOR value(5))/c
      PRINT time, x, y, z

    CASE ELSE
      PRINT "Read from "; characteristic.uuid

  END SELECT
ELSE IF kind = 3 AND err <> 0 THEN
  PRINT "Error writing "; characteristic.uuid; ": ("; err; ") "; message
END IF
END SUB
```

The Magnetometer

A magnetometer measures the strength of a magnetic field. The most common use is an electronic compass, but a magnetometer can also be used as simple metal detector (as we did in Chapter 3) or even a current detector, since electrons flowing through a wire also produce a magnetic field.

Since magnetic fields are measured in teslas, the units are:

$$T = \frac{Vs}{m^2}$$

or one volt-second per meter squared. The Earth's magnetic field varies from about 0.000025 to 0.000065 teslas, so it's very common to see magnetic field strength given in µT, or micro-teslas. It's just easier to deal with 25 to 65 µT.

The SensorTag uses the MAG3110 chip, which reports magnetic field strength from –1T to 1T along each of three axes. The chip claims a sensitivity of 0.1 µT, and the SensorTag reports the values to a precision of about 0.03 µT. Of course, that means any digits beyond one-tenth of a micro-tesla should be ignored.

Accessing the Magnetometer

Using the magnetometer on the SensorTag is almost identical to using the accelerometer. Looking at the Generic Attribute Profile (GATT) in Figure 6-9, you'll see three characteristics that exactly parallel the characteristics for the accelerometer.

0x3E	62	0x2800	GATT_PRIMARY_SERVICE_UUID	0xAA30 (MAGNETOMETER_SERV_UUID)	GATT_PERMIT_READ	Start of Sensor Profile Magnetometer Service
0x3F	63	0x2803	GATT_CHARACTER_UUID	12 (properties: read/notify) 40 00 (handle: 0x0040) 31 AA (UUID: 0xAA31)	GATT_PERMIT_READ	
0x40	64	0xAA31	MAGNETOMETER_DATA_UUID	00:00:00:00:00:00 (6 bytes)	GATT_PERMIT_READ	XMSB:XLSB:YMSB:YLSB:ZMSB:ZLSB Coordinates
0x41	65	0x2902	GATT_CLIENT_CHAR_CFG_UUID	00:00 (2 bytes)	GATT_PERMIT_READ | GATT_PERMIT_WRITE	Write "01:00" to enable notifications, "00:00" to disable
0x42	66	0x2901	GATT_CHAR_USER_DESC_UUID	"Mag. Data" (10 bytes)	GATT_PERMIT_READ	
0x43	67	0x2803	GATT_CHARACTER_UUID	0A (properties: read/write) 44 00 (handle: 0x0044) 32 AA (UUID: 0xAA32)	GATT_PERMIT_READ	
0x44	68	0xAA32	MAGNETOMETER_CONF_UUID	1 (1 byte)	GATT_PERMIT_READ | GATT_PERMIT_WRITE	Write "01" to start Sensor and Measurements, "00" to put to sleep
0x45	69	0x2901	GATT_CHAR_USER_DESC_UUID	"Mag. Conf." (11 bytes)	GATT_PERMIT_READ	
0x46	70	0x2803	GATT_CHARACTER_UUID	0A (properties: read/write) 47 00 (handle: 0x0047) 33 AA (UUID: 0xAA33)	GATT_PERMIT_READ	
0x47	71	0xAA33	MAGNETOMETER_PERI_UUID	1 (1 byte)	GATT_PERMIT_READ | GATT_PERMIT_WRITE	Period = [Input*10]ms, default 2000ms, lower limit 100 ms
0x48	72	0x2901	GATT_CHAR_USER_DESC_UUID	"Mag. Period" (12 bytes)	GATT_PERMIT_READ	

Figure 6-9. Magnetometer GATT profile, courtesy of Texas Instruments

The complete program for accessing the magnetometer is listed at the end of this section for reference. It's also included in techBASIC and techBASIC Sampler. Look for the app called SensorTag Magnetometer in the *O'Reilly Books* folder.

The SensorTag program from the start of the chapter also reads the magnetometer, displaying it using an interactive graph in a nice GUI interface. It's in the *O'Reilly Books* folder as TI SensorTag.

The AA32 characteristic is used to turn the magnetometer on and off. Our program just turns it on:

```
CASE "F000AA32" & servicesHeader
  ! Turn the magnetometer sensor on.
  IF debug THEN PRINT "Magnetometer on."
  DIM value(1) as INTEGER
  value(1) = 1
  peripheral.writeCharacteristic(characteristics(i), value, 1)
```

Pass a 0 instead of a 1 to turn the magnetometer off.

AA31 is used to read the magnetometer. There are two ways to read the magnetometer, either reading it when the program wants a magnetic field value or asking the magnetometer to notify the program whenever a new sensor reading is available. The first step is to tell the magnetometer which way we want to read values. To receive notifications when a value is available, write a 1 and a 0 to the characteristic, and then ask it to start sending notifications:

```
CASE "F000AA31" & servicesHeader
  ! Tell the magnetometer to begin sending data.
  IF debug THEN PRINT "Start magnetometer."
  DIM value(2) as INTEGER
  value = [0, 1]
  peripheral.writeCharacteristic(characteristics(i), value, 0)
  peripheral.setNotify(characteristics(i), 1)
```

To read a single value, write two zero bytes to this characteristic:

```
DIM value(2) as INTEGER
value = [0, 0]
peripheral.writeCharacteristic(characteristics(i), value, 0)
```

Then, whenever a value is needed, read the characteristic:

```
peripheral.readCharacteristic(characteristics(i), 1)
```

The magnetometer sends samples about once every two seconds unless you tell it otherwise. Use AA33 to set the sample rate. The value is expressed in tens of milliseconds, so passing 200 gives the default sample rate of once every two seconds. Keep in mind that the hardware uses the sample rate as a suggestion, not a firm value! If the time between samples is really important, record the time with the sample value.

Whether you read the data once using readCharacteristic or ask for notifications with setNotify, the values are always returned using a call to BLECharacteristicInfo.

The one difference between the GATT profiles for the accelerometer and magnetometer is that the magnetometer returns a two-byte value for each axis rather than the single-byte value returned by the accelerometer. Each value is a two-byte signed integer with the least significant byte first. The integers must be divided by 65,536 and multiplied by 2,000 to yield the final range of –1,000 to 1,000 µT. Here's the BASIC code to take care of reading and printing the values:

```
SUB BLECharacteristicInfo (time AS DOUBLE, _
                           peripheral AS BLEPeripheral, _
                           characteristic AS BLECharacteristic, _
                           kind AS INTEGER, _
                           message AS STRING, _
                           err AS LONG)
  IF kind = 2 THEN
    DIM value(1) AS INTEGER
```

```
      value = characteristic.value
      SELECT CASE characteristic.uuid
        CASE "F000AA31" & servicesHeader
          ! Update the magnetometer.
          c = 65536.0/2000.0
          x = ((value(2) << 8) BITOR value(1))/c
          y = ((value(4) << 8) BITOR value(3))/c
          z = ((value(6) << 8) BITOR value(5))/c
          PRINT time, x, y, z

        CASE ELSE
          PRINT "Read from "; characteristic.uuid

      END SELECT
    ELSE IF kind = 3 AND err <> 0 THEN
      PRINT "Error writing "; characteristic.uuid; ": ("; err; ") "; message
    END IF
    END SUB
```

Using the Magnetometer

Calibrating a magnetometer can be pretty tricky. We could use the Earth itself to calibrate the magnetometer, but that doesn't work so well, for two reasons. First, the magnetic field for the Earth is far from constant. In fact, there are places on the Earth where a compass points south, not north, and others where a compass won't work at all! The Earth's magnetic field is also not parallel to the surface of the Earth. The angle between horizontal and the actual direction of the magnetic field is called the *inclination* and is measured with a device called a *dip circle*—essentially, a compass turned on its side. Since compasses are so ubiquitous for navigation, these values have been mapped pretty carefully. You can get rough values from maps like the ones in the Wikipedia article on the Earth's magnetic field (*http://en.wikipedia.org/wiki/Earth%27s_magnetic_field*).

With the inclination, declination (the difference between magnetic north and true north), and field strength in hand, you have a pretty good source for calibrating your magnetometer.

Of course, you're going to move the SensorTag well away from refrigerator magnets, electric motors, extension cords, metal tables, and the like before calibration, right? And what about the SensorTag itself? Yes, the SensorTag is also generating magnetic fields, and they will even vary depending on which sensors you turn on or off! That really makes the calibration job difficult.

Fortunately, you may not need accurate calibration. If you want to measure the precise magnetic field from an electric motor, by all means, go to the trouble of calibrating the device accurately. If you want to know the rough direction of north, it's good enough to map your magnetometer readings to the values returned by a compass. For some applications, like detecting a current in a wire running behind a wall, you don't need to calibrate the device at all, since you're really just watching for the magnetic field to change.

Here are a few lines of sample output from the magnetometer:

```
385593559.248537003994      71.350098      -104.705803      -9.063721
385593560.098580002785      58.654781      -93.872063       -6.072998
385593563.048699021339      67.016602      -95.733635       -23.895262
385593563.898690998554      58.685299      -96.191399       -4.394531
```

The Source

Here's the source for the magnetometer program. It prints the time when the sample was returned and the magnetic field strength along each axis to the console, updating the values about once a second. This is a great program to experiment a bit with the magnetometer. The magnetometer is one of the rare sensors where I get more from the plot of the strength than from the text values, so be sure and try the SensorTag program, too.

```
! Simple program to access the humidity sensor on the TI SensorTag.

! Set up variables to hold the peripheral and the characteristics
! for the battery and buzzer.
DIM sensorTag AS BLEPeripheral

! We will look for these services.
DIM servicesHeader AS STRING, services(1) AS STRING
servicesHeader = "-0451-4000-B000-000000000000"
services(1) = "F000AA30" & servicesHeader : ! Magnetometer
mag% = 1

! Start the BLE service and begin scanning for devices.
debug = 0
BLE.startBLE
DIM uuid(0) AS STRING
BLE.startScan(uuid)

! Called when a peripheral is found. If it is a Sensor Tag, we
! initiate a connection to it and stop scanning for peripherals.
!
! Parameters:
!    time - The time when the peripheral was discovered.
!    peripheral - The peripheral that was discovered.
!    services - List of services offered by the device.
!    advertisements - Advertisements (information provided by the
!        device without the need to read a service/characteristic)
!    rssi - Received Signal Strength Indicator
!
SUB BLEDiscoveredPeripheral (time AS DOUBLE, _
                             peripheral AS BLEPeripheral, _
                             services() AS STRING, _
```

```
                                    advertisements(,) AS STRING, _
                                    rssi)
    IF peripheral.bleName = "TI BLE Sensor Tag" THEN
      sensorTag = peripheral
      BLE.connect(sensorTag)
      BLE.stopScan
    END IF
    END SUB


    ! Called to report information about the connection status of the
    ! peripheral or to report that services have been discovered.
    !
    ! Parameters:
    !     time - The time when the information was received.
    !     peripheral - The peripheral.
    !     kind - The kind of call. One of
    !         1 - Connection completed
    !         2 - Connection failed
    !         3 - Connection lost
    !         4 - Services discovered
    !     message - For errors, a human-readable error message.
    !     err - If there was an error, the Apple error number. If there
    !         was no error, this value is 0.
    !
    SUB BLEPeripheralInfo (time AS DOUBLE, _
                           peripheral AS BLEPeripheral, _
                           kind AS INTEGER, _
                           message AS STRING, _
                           err AS LONG)
    IF kind = 1 THEN
      ! The connection was established. Look for available services.
      IF debug THEN PRINT "Connection made."
      peripheral.discoverServices(uuid)
    ELSE IF kind = 2 OR kind = 3 THEN
      IF debug THEN PRINT "Connection lost: "; kind
      BLE.connect(sensorTag)
    ELSE IF kind = 4 THEN
      ! Services were found. If it is one of the ones we are interested
      ! in, begin discovery of its characteristics.
      DIM availableServices(1) AS BLEService
      availableServices = peripheral.services
      FOR s = 1 to UBOUND(services, 1)
        FOR a = 1 TO UBOUND(availableServices, 1)
          IF services(s) = availableServices(a).uuid THEN
            IF debug THEN PRINT "Discovering characteristics for "; services(s)
            peripheral.discoverCharacteristics(uuid, availableServices(a))
          END IF
        NEXT
      NEXT
    END IF
    END SUB
```

```
! Called to report information about a characteristic or included
! services for a service. If it is one we are interested in, start
! handling it.
!
! Parameters:
!    time - The time when the information was received.
!    peripheral - The peripheral.
!    service - The service whose characteristic or included
!        service was found.
!    kind - The kind of call. One of
!        1 - Characteristics found
!        2 - Included services found
!    message - For errors, a human-readable error message.
!    err - If there was an error, the Apple error number. If there
!        was no error, this value is 0.
!
SUB BLEServiceInfo (time AS DOUBLE, _
                    peripheral AS BLEPeripheral, _
                    service AS BLEService, _
                    kind AS INTEGER, _
                    message AS STRING, _
                    err AS LONG)
IF kind = 1 THEN
  ! Get the characteristics.
  DIM characteristics(1) AS BLECharacteristic
  characteristics = service.characteristics
  FOR i = 1 TO UBOUND(characteristics, 1)
    IF service.uuid = services(mag%) THEN
        ! Found the magnetometer.
        SELECT CASE characteristics(i).uuid
          CASE "F000AA31" & servicesHeader
            ! Tell the magnetometer to begin sending data.
            IF debug THEN PRINT "Start magnetometer."
            DIM value(2) as INTEGER
            value = [0, 1]
            peripheral.writeCharacteristic(characteristics(i), value, 0)
            peripheral.setNotify(characteristics(i), 1)

          CASE "F000AA32" & servicesHeader
            ! Turn the magnetometer sensor on.
            IF debug THEN PRINT "Magnetometer on."
            DIM value(1) as INTEGER
            value(1) = 1
            peripheral.writeCharacteristic(characteristics(i), value, 1)

          CASE "F000AA33" & servicesHeader
            ! Set the sample rate to 100ms.
            DIM value(1) as INTEGER
            value(1) = 100
            IF debug THEN PRINT "Setting magnetometer sample rate to "; value(1)
            peripheral.writeCharacteristic(characteristics(i), value, 1)
```

```
      END SELECT
    END IF
  NEXT
END IF
END SUB


! Called to return information from a characteristic.
!
! Parameters:
!    time - The time when the information was received.
!    peripheral - The peripheral.
!    characteristic - The characteristic whose information
!        changed.
!    kind - The kind of call. One of
!        1 - Called after a discoverDescriptors call.
!        2 - Called after a readCharacteristics call.
!        3 - Called to report status after a writeCharacteristics
!            call.
!    message - For errors, a human-readable error message.
!    err - If there was an error, the Apple error number. If there
!        was no error, this value is 0.
!
SUB BLECharacteristicInfo (time AS DOUBLE, _
                           peripheral AS BLEPeripheral, _
                           characteristic AS BLECharacteristic, _
                           kind AS INTEGER, _
                           message AS STRING, _
                           err AS LONG)
IF kind = 2 THEN
  DIM value(1) AS INTEGER
  value = characteristic.value
  SELECT CASE characteristic.uuid
    CASE "F000AA31" & servicesHeader
      ! Update the magnetometer.
      c = 65536.0/2000.0
      x = ((value(2) << 8) BITOR value(1))/c
      y = ((value(4) << 8) BITOR value(3))/c
      z = ((value(6) << 8) BITOR value(5))/c
      PRINT time, x, y, z

    CASE ELSE
      PRINT "Read from "; characteristic.uuid

  END SELECT
ELSE IF kind = 3 AND err <> 0 THEN
  PRINT "Error writing "; characteristic.uuid; ": ("; err; ") "; message
END IF
END SUB
```

The Humidity Sensor (Hygrometer)

A hygrometer measures the moisture in an environment—in our case, the humidity in air. We all know what humidity is (especially those of us who have lived in the Ohio River Valley). It's all that water in the air that doesn't seem to be falling as rain—yet! But what is it really, and what does a humidity meter measure?

First, humidity is not clouds or fog. The clouds you see are well beyond humidity; they are tiny drops of water or ice crystals suspended in air, like cocoa powder suspended in water in your hot chocolate. Humidity is the water that's actually part of the air, existing as a gas of individual molecules mixed with the nitrogen, oxygen, carbon dioxide, and other gases in the air. It's more like sugar dissolved in water than cocoa particles suspended in it. As you know from experience, though, there is only so much sugar a glass of water can hold. If you put too much sugar in a glass of water, no amount of stirring will dissolve it. The same thing happens with air—it can only hold a certain amount of water. The amount varies with temperature and pressure, but there is always a limit. That limit is the saturation amount, known as the *saturated vapor pressure*. The relative humidity is a measure of the percentage of water that is in the air compared to the amount of water the air can hold. The Ohio River Valley in the summer often has very high humidity, so sweat doesn't evaporate well, causing us to feel very hot. Here in Albuquerque, we have a dry heat. As I write this, the humidity is 15%. It really does make a difference—sweat evaporates quickly, cooling the body, so a 90° day doesn't really feel as bad as 80° in a humid environment.

The SHT21 Digital Humidity Sensor used on the SensorTag reports relative humidity. It needs the temperature to calculate the humidity, so it reports that, too.

Accessing the Hygrometer

Figure 6-10 shows the Generic Attribute Profile (GATT) for the humidity service.

0x36	54	0x2800	GATT_PRIMARY_SERVICE_UUID	0xAA20 (HUMIDITY_SERV_UUID)	GATT_PERMIT_READ	Start of Sensor Profile Humidty Service
0x37	55	0x2803	GATT_CHARACTER_UUID	12 (properties: read/notify) 36 00 (handle: 0x0036) 21 AA (UUID: 0xAA21)	GATT_PERMIT_READ	
0x38	56	0xAA21	HUMIDITY_DATA_UUID	00:00:00:00 (4 bytes)	GATT_PERMIT_READ	TempLSB:TempMSB:HumidityLSB:HumidityMSB
0x39	57	0x2902	GATT_CLIENT_CHAR_CFG_UUID	00:00 (2 bytes)	GATT_PERMIT_READ | GATT_PERMIT_WRITE	Write "01:00" to enable notifications
0x3A	58	0x2901	GATT_CHAR_USER_DESC_UUID	"Humid. Data" (14 bytes)	GATT_PERMIT_READ	
0x3B	59	0x2803	GATT_CHARACTER_UUID	0A (properties: read/write) 3C 00 (handle: 0x003C) 22 AA (UUID: 0xAA22)	GATT_PERMIT_READ	
0x3C	60	0xAA22	HUMIDITY_CONF_UUID	1 (1 byte)	GATT_PERMIT_READ | GATT_PERMIT_WRITE	Write "01" to start Sensor and Measurements, "00" to put to sleep
0x3D	61	0x2901	GATT_CHAR_USER_DESC_UUID	"Humid. Conf." (15 bytes)	GATT_PERMIT_READ	

Figure 6-10. Hygrometer GATT profile, courtesy of Texas Instruments

The complete program for accessing the hygrometer is listed at the end of this section for reference. It's also included in techBASIC and tech-BASIC Sampler. Look for the app called SensorTag Humidity in the *O'Reilly Books* folder.

The SensorTag program from the start of the chapter also reads the hygrometer, displaying it using an interactive graph in a nice GUI interface. It's in the *O'Reilly Books* folder as TI SensorTag.

Attribute AA22 turns the sensor on or off. Our program turns it on by writing a 1 to this attribute. You can turn it off by writing a 0, but this program doesn't turn the sensor off:

```
CASE "F000AA22" & servicesHeader
  DIM value(1) as INTEGER
  value(1) = 1
  peripheral.writeCharacteristic(characteristics(i), value, 1)
```

AA21 is used to read the humidity and temperature. There are two ways to read the hygrometer, either reading it when the program wants the humidity or asking the hygrometer to notify the program whenever a new sensor reading is available. The first step is to tell the hygrometer which way we want to read values. Write a 1 and a 0 to the characteristic to receive notifications when a value is available, then ask it to start sending notifications:

```
CASE "F000AA22" & servicesHeader
  ! Turn the humidity sensor on.
  IF debug THEN PRINT "Humidity on."
  DIM value(1) as INTEGER
  value(1) = 1
  peripheral.writeCharacteristic(characteristics(i), value, 1)
```

To read a single value, write two zero bytes to this characteristic:

```
DIM value(2) as INTEGER
value = [0, 0]
peripheral.writeCharacteristic(characteristics(i), value, 0)
```

Then, whenever a value is needed, read the characteristic:

```
peripheral.readCharacteristic(characteristics(i), 1)
```

Whether you read the data once using readCharacteristic or ask for notifications with setNotify, the values are always returned using a call to BLECharacteristicInfo. The temperature and humidity are returned in two unsigned integers, with the temperature in the first two bytes and the humidity in the second. Both are unsigned values stored least significant byte first. Here's the conversion from the value returned to a relative humidity:

$$h = -6 + 125\,h_{3,4}/65536$$

where $h_{3,4}$ is the value returned from the SensorTag and h is the relative humidity, which should have a value from 0 to 100. While it is possible to have a humidity over 100%, a situation known as super-saturation, the SHT21 sensor doesn't claim to work in that region, so you should not see values over 100.

Use this formula to convert the raw temperature values to degrees Celsius:

$$t = -46.86 + 175.72\,t_{1,2}/65536$$

Here's what we get after converting the equations to BASIC:

```
SUB BLECharacteristicInfo (time AS DOUBLE, _
                           peripheral AS BLEPeripheral, _
                           characteristic AS BLECharacteristic, _
                           kind AS INTEGER, _
                           message AS STRING, _
                           err AS LONG)
IF kind = 2 THEN
  DIM value(1) AS INTEGER
  value = characteristic.value
  SELECT CASE characteristic.uuid
    CASE "F000AA21" & servicesHeader
      ! Update the humidity indicator.
      v& = value(3) BITOR (value(4) << 8)
      v& = v& BITAND $00FFFC
      v = v&*125.0/65536 - 6.0

      t& = (value(1) BITOR (value(2) << 8)) BITAND $00FFFF
      t = -46.85 + 175.72*t&/65536.0
      PRINT "Relative humidity: "; v; ", temperature = "; t

    CASE ELSE
      PRINT "Read from "; characteristic.uuid

  END SELECT
ELSE IF kind = 3 AND err <> 0 THEN
  PRINT "Error writing "; characteristic.uuid; ": ("; err; ") "; message
END IF
END SUB
```

Running the program gives output like this:

```
Relative humidity: 15.057129, temperature = 24.075073
Relative humidity: 15.209717, temperature = 24.075073
Relative humidity: 15.324158, temperature = 24.075073
```

The Source

Here's the complete source for the hygrometer program:

```
! Simple program to access the humidity sensor on the TI SensorTag.
```

```
! Set up variables to hold the peripheral and the characteristics
! for the battery and buzzer.
DIM sensorTag AS BLEPeripheral

! We will look for these services.
DIM servicesHeader AS STRING, services(1) AS STRING
servicesHeader = "-0451-4000-B000-000000000000"
services(1) = "F000AA20" & servicesHeader : ! Humidity
hum% = 1

! Start the BLE service and begin scanning for devices.
debug = 0
BLE.startBLE
DIM uuid(0) AS STRING
BLE.startScan(uuid)

! Called when a peripheral is found. If it is a Sensor Tag, we
! initiate a connection to it and stop scanning for peripherals.
!
! Parameters:
!    time - The time when the peripheral was discovered.
!    peripheral - The peripheral that was discovered.
!    services - List of services offered by the device.
!    advertisements - Advertisements (information provided by the
!        device without the need to read a service/characteristic)
!    rssi - Received Signal Strength Indicator
!
SUB BLEDiscoveredPeripheral (time AS DOUBLE, _
                            peripheral AS BLEPeripheral, _
                            services() AS STRING, _
                            advertisements(,) AS STRING, _
                            rssi)
IF peripheral.bleName = "TI BLE Sensor Tag" THEN
  sensorTag = peripheral
  BLE.connect(sensorTag)
  BLE.stopScan
END IF
END SUB

! Called to report information about the connection status of the
! peripheral or to report that services have been discovered.
!
! Parameters:
!    time - The time when the information was received.
!    peripheral - The peripheral.
!    kind - The kind of call. One of
!        1 - Connection completed
!        2 - Connection failed
!        3 - Connection lost
!        4 - Services discovered
!    message - For errors, a human-readable error message.
!    err - If there was an error, the Apple error number. If there
```

```
!        was no error, this value is 0.
!
SUB BLEPeripheralInfo (time AS DOUBLE, _
                       peripheral AS BLEPeripheral, _
                       kind AS INTEGER, _
                       message AS STRING, _
                       err AS LONG)
IF kind = 1 THEN
  ! The connection was established. Look for available services.
  IF debug THEN PRINT "Connection made."
  peripheral.discoverServices(uuid)
ELSE IF kind = 2 OR kind = 3 THEN
  IF debug THEN PRINT "Connection lost: "; kind
  BLE.connect(sensorTag)
ELSE IF kind = 4 THEN
  ! Services were found. If it is one of the ones we are interested
  ! in, begin discovery of its characteristics.
  DIM availableServices(1) AS BLEService
  availableServices = peripheral.services
  FOR s = 1 to UBOUND(services, 1)
    FOR a = 1 TO UBOUND(availableServices, 1)
      IF services(s) = availableServices(a).uuid THEN
        IF debug THEN PRINT "Discovering characteristics for "; services(s)
        peripheral.discoverCharacteristics(uuid, availableServices(a))
      END IF
    NEXT
  NEXT
END IF
END SUB


! Called to report information about a characteristic or included
! services for a service. If it is one we are interested in, start
! handling it.
!
! Parameters:
!    time - The time when the information was received.
!    peripheral - The peripheral.
!    service - The service whose characteristic or included
!        service was found.
!    kind - The kind of call. One of
!        1 - Characteristics found
!        2 - Included services found
!    message - For errors, a human-readable error message.
!    err - If there was an error, the Apple error number. If there
!        was no error, this value is 0.
!
SUB BLEServiceInfo (time AS DOUBLE, _
                    peripheral AS BLEPeripheral, _
                    service AS BLEService, _
                    kind AS INTEGER, _
                    message AS STRING, _
```

```
                            err AS LONG)
IF kind = 1 THEN
  ! Get the characteristics.
  DIM characteristics(1) AS BLECharacteristic
  characteristics = service.characteristics
  FOR i = 1 TO UBOUND(characteristics, 1)
    IF service.uuid = services(hum%) THEN
      ! Found the humidity sensor.
      SELECT CASE characteristics(i).uuid
        CASE "F000AA21" & servicesHeader
          ! Tell the humidity sensor to begin sending data.
          IF debug THEN PRINT "Start humidity sensor."
          DIM value(2) as INTEGER
          value = [0, 1]
          peripheral.writeCharacteristic(characteristics(i), value, 0)
          peripheral.setNotify(characteristics(i), 1)

        CASE "F000AA22" & servicesHeader
          ! Turn the humidity sensor on.
          IF debug THEN PRINT "Humidity on."
          DIM value(1) as INTEGER
          value(1) = 1
          peripheral.writeCharacteristic(characteristics(i), value, 1)
      END SELECT
    END IF
  NEXT
END IF
END SUB


! Called to return information from a characteristic.
!
! Parameters:
!    time - The time when the information was received.
!    peripheral - The peripheral.
!    characteristic - The characteristic whose information
!        changed.
!    kind - The kind of call. One of
!        1 - Called after a discoverDescriptors call.
!        2 - Called after a readCharacteristics call.
!        3 - Called to report status after a writeCharacteristics
!            call.
!    message - For errors, a human-readable error message.
!    err - If there was an error, the Apple error number. If there
!        was no error, this value is 0.
!
SUB BLECharacteristicInfo (time AS DOUBLE, _
                           peripheral AS BLEPeripheral, _
                           characteristic AS BLECharacteristic, _
                           kind AS INTEGER, _
                           message AS STRING, _
                           err AS LONG)
```

```
IF kind = 2 THEN
  DIM value(1) AS INTEGER
  value = characteristic.value
  SELECT CASE characteristic.uuid
    CASE "F000AA21" & servicesHeader
      ! Update the humidity indicator.
      v& = value(3) BITOR (value(4) << 8)
      v& = v& BITAND $00FFFC
      v = v&*125.0/65536 - 6.0

      t& = (value(1) BITOR (value(2) << 8)) BITAND $00FFFF
      t = -46.85 + 175.72*t&/65536.0
      PRINT "Relative humidity: "; v; ", temperature = "; t

    CASE ELSE
      PRINT "Read from "; characteristic.uuid

  END SELECT
ELSE IF kind = 3 AND err <> 0 THEN
  PRINT "Error writing "; characteristic.uuid; ": ("; err; ") "; message
END IF
END SUB
```

The Thermometer

The last of the six sensors in the SensorTag is a thermometer. The SensorTag uses the TMP006 chip to read temperature. The amazing thing about this chip is not that it can sense temperature. Heck, a thermocouple can do that. The amazing thing is that it can sense temperature remotely by examining the infrared heat signature from an object. The thermometer sensor actually returns two temperatures. One is the temperature on the circuit board, called the *die temperature*. The other is the temperature of the object the thermometer is looking at through the window on the front of the SensorTag. This is called the *target temperature*.

Accessing the Thermometer

Figure 6-11 shows the Generic Attribute Profile (GATT) for the thermometer service. It's pretty simple, with just two attributes.

0x23	35	0x2800	GATT_PRIMARY_SERVICE_UUID	**0xAA00** (IRTEMPERATURE_SERV_UUID)	GATT_PERMIT_READ	Start of Sensor Profile Temperature Service
0x24	36	0x2803	GATT_CHARACTER_UUID	12 (properties: read/notify) 25 00 (handle: 0x0025) 01 AA (UUID: **0xAA01**)	GATT_PERMIT_READ	
0x25	37	**0xAA01**	IRTEMPERATURE_DATA_UUID	00:00:00:00 (4 bytes)	GATT_PERMIT_READ	ObjectLSB:ObjectMSB:AmbientLSB:AmbientMSB
0x26	38	0x2902	GATT_CLIENT_CHAR_CFG_UUID	00:00 (2 bytes)	GATT_PERMIT_READ | GATT_PERMIT_WRITE	Write "01:00" to enable notifications, "00:00" to disable
0x27	39	0x2901	GATT_CHAR_USER_DESC_UUID	"IR Temp. Data" (14 bytes)	GATT_PERMIT_READ	
0x28	40	0x2803	GATT_CHARACTER_UUID	0A (properties: read/write) 29 00 (handle: 0x0029) 02AA (UUID: **0xAA02**)	GATT_PERMIT_READ	
0x29	41	**0xAA02**	IRTEMPERATURE_CONF_UUID	1 (1 byte)	GATT_PERMIT_READ | GATT_PERMIT_WRITE	Write "01" to start Sensor and Measurements, "00" to put to sleep
0x2A	42	0x2901	GATT_CHAR_USER_DESC_UUID	"IR Temp. Conf." (15 bytes)	GATT_PERMIT_READ	

Figure 6-11. Thermometer GATT profile, courtesy of Texas Instruments

Attribute 0xAA02 turns the sensor on or off. Our program turns it on by writing a 1 to this attribute. You can turn it off by writing a 0, but this simple program doesn't turn the sensor off:

```
CASE "F000AA02" & servicesHeader
  ! Turn the thermometer sensor on.
  IF debug THEN PRINT "Thermometer on."
  DIM value(1) as INTEGER
  value(1) = 1
  peripheral.writeCharacteristic(characteristics(i), value, 1)
```

0xAA01 is used to read the thermometer. As with all of the other sensors, there are two ways to do this, either reading it when the program wants the temperature or asking the thermometer to notify the program whenever a new sensor reading is available. The first step is to tell the thermometer which way we want to read values. Write a 1 and a 0 to the characteristic to receive notifications when a value is available, then ask it to start sending notifications:

```
CASE "F000AA01" & servicesHeader
  ! Tell the thermometer to begin sending data.
  IF debug THEN PRINT "Start thermometer."
  DIM value(2) as INTEGER
  value = [0, 1]
  peripheral.writeCharacteristic(characteristics(i), value, 0)
  peripheral.setNotify(characteristics(i), 1)
```

As with the other sensors, values are returned using a call to BLECharacteristicInfo. The returned value has two different temperatures stored in four bytes. Bytes 3 and 4 are the die temperature, while bytes 1 and 2 contain the raw data used to calculate the target temperature. Both are unsigned values stored least significant byte first. The conversion to degrees Celsius is:

$$t_{die} = t_{3,4}/128$$

$$V_{obj} = 1.5625^{-7}t_{1,2}$$

$$t_{die2} = t_{die} + 273.15$$

$$S = 6.4^{-14}(1 + 1.75^{-3}(t_{die2} - 298.15) - 1.678^{-5}(t_{die2} - 298.15)^2)$$

$$V_{os} = -2.9^{-5} - 5.7^{-7}(t_{die2} - 298.15) + (t_{die2} - 298.15)^2$$

$$N = V_{obj} - V_{os} + 13.4 \cdot (V_{obj} - V_{os})^2$$

$$t_{target} = (t_{die2}^4 + N/S)^{0.25} - 273.15$$

where $t_{1,2}$ is the raw value returned in the first two bytes, and $t_{3,4}$ is the raw value returned in the second pair of bytes. Here's what we get after converting the equations to BASIC:

```
SUB BLECharacteristicInfo (time AS DOUBLE, _
                           peripheral AS BLEPeripheral, _
                           characteristic AS BLECharacteristic, _
                           kind AS INTEGER, _
                           message AS STRING, _
                           err AS LONG)
IF kind = 2 THEN
 DIM value(1) AS INTEGER
 value = characteristic.value
 SELECT CASE characteristic.uuid
   CASE "F000AA01" & servicesHeader
     ! Update the thermometer.
     temp = value(3) BITOR (value(4) << 8)
     temp = temp/128.0

     target = value(1) BITOR (value(2) << 8)
     target = target*0.00000015625
     die2 = 273.15 + temp
     s0 = 6.4e-14
     a1 = 1.75e-3
     a2 = -1.678e-5
     b0 = -2.9e-5
     b1 = -5.7e-7
     b2 = 4.63e-9
     c2 = 13.4
     tref = 298.15
     dt2 = (die2 - tref)*(dies2 - tref)
     S = s0*(1 + a1*(die2 - tref) + a2*dt2)
     Vos = b0 + b1*(die2 - tref) + b2*dt2
     fObj = (target - Vos) + c2*((target - Vos)*(target - Vos))
     tObj = (die2^4 + fObj/S)^0.25
     tObj = tObj - 273.15
     if Math.isNaN(tObj) then
        tObj = 1
     end if

     PRINT "Die temperature: "; temp; "  Target temperature: "; tObj
```

```
    CASE ELSE
        PRINT "Read from "; characteristic.uuid

    END SELECT
    ELSE IF kind = 3 AND err <> 0 THEN
      PRINT "Error writing "; characteristic.uuid; ": ("; err; ") "; message
    END IF
END SUB
```

Using the Thermometer

The target temperature tends to jump around a bit as the chip sees various temperatures. Here's the output from the program with the target temperature reading the inside of my mouth—the temperature should be about 37, since the reading is in Celsius:

```
Die temperature: 23.75  Target temperature: 64.343353
Die temperature: 23.75  Target temperature: 28.774109
Die temperature: 23.75  Target temperature: 33.092346
Die temperature: 23.78125  Target temperature: 40.621277
```

I'd suggest keeping a running average of the last few values returned to keep this stable. It will make the thermometer slower to respond to a change, but it will also give a steadier reading that is less inclined to jump around due to noise.

Of course, if you're after the temperature at the SensorTag, you can just use the die temperature. This would be fine for a weather station. The SensorTag doesn't use much power, so it won't heat the thermometer much.

The Source

Here's the complete source for the thermometer program. As usual, it's in the *O'Reilly Books* folder in techBASIC and techBASIC Sampler. This one is called SensorTag Thermometer:

```
! Simple program to access the thermometer on the TI SensorTag.

! Set up variables to hold the peripheral and the characteristics
! for the battery and buzzer.
DIM sensorTag AS BLEPeripheral

! We will look for these services.
DIM servicesHeader AS STRING, services(1) AS STRING
servicesHeader = "-0451-4000-B000-000000000000"
services(1) = "F000AA00" & servicesHeader : ! Thermometer
therm% = 1

! Start the BLE service and begin scanning for devices.
debug = 0
BLE.startBLE
DIM uuid(0) AS STRING
BLE.startScan(uuid)
```

```
! Called when a peripheral is found. If it is a Sensor Tag, we
! initiate a connection to it and stop scanning for peripherals.
!
! Parameters:
!    time - The time when the peripheral was discovered.
!    peripheral - The peripheral that was discovered.
!    services - List of services offered by the device.
!    advertisements - Advertisements (information provided by the
!        device without the need to read a service/characteristic)
!    rssi - Received Signal Strength Indicator
!
SUB BLEDiscoveredPeripheral (time AS DOUBLE, _
                              peripheral AS BLEPeripheral, _
                              services() AS STRING, _
                              advertisements(,) AS STRING, _
                              rssi)
IF peripheral.bleName = "TI BLE Sensor Tag" THEN
 sensorTag = peripheral
 BLE.connect(sensorTag)
 BLE.stopScan
END IF
END SUB


! Called to report information about the connection status of the
! peripheral or to report that services have been discovered.
!
! Parameters:
!    time - The time when the information was received.
!    peripheral - The peripheral.
!    kind - The kind of call. One of
!        1 - Connection completed
!        2 - Connection failed
!        3 - Connection lost
!        4 - Services discovered
!    message - For errors, a human-readable error message.
!    err - If there was an error, the Apple error number. If there
!        was no error, this value is 0.
!
SUB BLEPeripheralInfo (time AS DOUBLE, _
                        peripheral AS BLEPeripheral, _
                        kind AS INTEGER, _
                        message AS STRING, _
                        err AS LONG)
IF kind = 1 THEN
 ! The connection was established. Look for available services.
 IF debug THEN PRINT "Connection made."
 peripheral.discoverServices(uuid)
ELSE IF kind = 2 OR kind = 3 THEN
 IF debug THEN PRINT "Connection lost: "; kind
 BLE.connect(sensorTag)
ELSE IF kind = 4 THEN
```

```
    ! Services were found. If it is one of the ones we are interested
    ! in, begin discovery of its characteristics.
    DIM availableServices(1) AS BLEService
    availableServices = peripheral.services
    FOR s = 1 to UBOUND(services, 1)
      FOR a = 1 TO UBOUND(availableServices, 1)
        IF services(s) = availableServices(a).uuid THEN
          IF debug THEN PRINT "Discovering characteristics for "; services(s)
          peripheral.discoverCharacteristics(uuid, availableServices(a))
        END IF
      NEXT
    NEXT
  END IF
END SUB


! Called to report information about a characteristic or included
! services for a service. If it is one we are interested in, start
! handling it.
!
! Parameters:
!    time - The time when the information was received.
!    peripheral - The peripheral.
!    service - The service whose characteristic or included
!       service was found.
!    kind - The kind of call. One of
!        1 - Characteristics found
!        2 - Included services found
!    message - For errors, a human-readable error message.
!    err - If there was an error, the Apple error number. If there
!       was no error, this value is 0.
!
SUB BLEServiceInfo (time AS DOUBLE, _
                    peripheral AS BLEPeripheral, _
                    service AS BLEService, _
                    kind AS INTEGER, _
                    message AS STRING, _
                    err AS LONG)
  IF kind = 1 THEN
    ! Get the characteristics.
    DIM characteristics(1) AS BLECharacteristic
    characteristics = service.characteristics
    FOR i = 1 TO UBOUND(characteristics, 1)
      IF service.uuid = services(therm%) THEN
        ! Found the thermometer.
        SELECT CASE characteristics(i).uuid
          CASE "F000AA01" & servicesHeader
            ! Tell the thermometer to begin sending data.
            IF debug THEN PRINT "Start thermometer."
            DIM value(2) as INTEGER
            value = [0, 1]
            peripheral.writeCharacteristic(characteristics(i), value, 0)
```

```
            peripheral.setNotify(characteristics(i), 1)

        CASE "F000AA02" & servicesHeader
          ! Turn the thermometer sensor on.
          IF debug THEN PRINT "Thermometer on."
          DIM value(1) as INTEGER
          value(1) = 1
          peripheral.writeCharacteristic(characteristics(i), value, 1)
      END SELECT
    END IF
  NEXT
 END IF
END IF
END SUB



! Called to return information from a characteristic.
!
! Parameters:
!     time - The time when the information was received.
!     peripheral - The peripheral.
!     characteristic - The characteristic whose information
!         changed.
!     kind - The kind of call. One of
!         1 - Called after a discoverDescriptors call.
!         2 - Called after a readCharacteristics call.
!         3 - Called to report status after a writeCharacteristics
!             call.
!     message - For errors, a human-readable error message.
!     err - If there was an error, the Apple error number. If there
!         was no error, this value is 0.
!
SUB BLECharacteristicInfo (time AS DOUBLE, _
                            peripheral AS BLEPeripheral, _
                            characteristic AS BLECharacteristic, _
                            kind AS INTEGER, _
                            message AS STRING, _
                            err AS LONG)
IF kind = 2 THEN
 DIM value(1) AS INTEGER
 value = characteristic.value
 SELECT CASE characteristic.uuid
   CASE "F000AA01" & servicesHeader
     ! Update the thermometer.
     temp = value(3) BITOR (value(4) << 8)
     temp = temp/128.0

     target = value(1) BITOR (value(2) << 8)
     target = target*0.00000015625
     die2 = 273.15 + temp
     s0 = 6.4e-14
     a1 = 1.75e-3
     a2 = -1.678e-5
```

```
b0 = -2.9e-5
b1 = -5.7e-7
b2 = 4.63e-9
c2 = 13.4
tref = 298.15
dt2 = (die2 - tref)*(dies2 - tref)
S = s0*(1 + a1*(die2 - tref) + a2*dt2)
Vos = b0 + b1*(die2 - tref) + b2*dt2
fObj = (target - Vos) + c2*((target - Vos)*(target - Vos))
tObj = (die2^4 + fObj/S)^0.25
tObj = tObj - 273.15
if Math.isNaN(tObj) then
  tObj = 1
end if

PRINT "Die temperature: "; temp; "  Target temperature: "; tObj
CASE ELSE
  PRINT "Read from "; characteristic.uuid

END SELECT
ELSE IF kind = 3 AND err <> 0 THEN
 PRINT "Error writing "; characteristic.uuid; ": ("; err; ") "; message
END IF
END SUB
```

Further Explorations

The programs in this chapter show how to access Bluetooth low energy devices and cover the major commands and techniques used for all such devices. There's more to learn, though. See the techBASIC Reference Manual for a host of other Bluetooth low energy commands, as well as several complete sample programs. Check the index or search the PDF for "Bluetooth low energy."

CHAPTER 7

Bluetooth Low Energy iPhone Rocket

About This Chapter

Prerequisites

Read Chapter 6 first to get a basic understanding of Bluetooth low energy technology and the Texas Instruments SensorTag used in this chapter.

Equipment

You will need an iPhone 4S or later or an iPod 5th Gen or later, running iOS 5 or later; a Texas Instruments SensorTag with the CC Debugger; and various parts for building the rockets. An iPad is best for analyzing the data, and can be used with the smaller rocket to collect data. See Tables 7-1, 7-2, and 7-3 for detailed parts lists. You will also need access to a Windows computer to update the SensorTag firmware.

Software

You will need a copy of techBASIC or techBASIC Sampler, the Flash Programmer, and the 8G firmware for the SensorTag. The Flash Programmer and 8G firmware are free downloads.

What You Will Learn

This project shows how to use a SensorTag as a sensor platform. It is used in two rockets, one that carries just the SensorTag and one that carries both the SensorTag and an iPhone as a recording device.

A Bit of Rocket Science

It was a beautiful fall morning as I carefully packed the parachute, slid in the engine, and installed the igniter in my model rocket. I started the data collection program and slid the payload with a TI Bluetooth low energy SensorTag and an iPhone 4S into the payload bay, as shown in Figure 7-1.

Figure 7-1. The ST-2 rocket, carrying an iPhone and a TI SensorTag

Yes, an iPhone.

My wife's iPhone.

Gulp.

Silently, I recited the astronaut's prayer, "Dear Lord, please don't let me screw up." Then I pushed the launch button.

The rocket lifted off smoothly, arcing into the air. The parachute deployed. Landing broke off a fin, but pulling the payload out, I saw the iPhone was still working, still collecting acceleration, rotation, and pressure readings!

We launched three more times that morning, once more with the same rocket after a quick field repair, and twice with another rocket that carried the SensorTag without an iPhone, transmitting the data using Bluetooth low energy to an iPhone held safely on the ground.

This chapter shows how it's done. We will use the ideas developed in Chapter 6 for accessing Bluetooth low energy devices from iOS. You'll get plans to build the rockets, programs to collect and analyze data, and even the data from my rocket flights.

The project in this chapter is both challenging and fun. It combines electronics, Bluetooth low energy communications, software, and the construction and flight of model rockets. Depending on your interests, each of these areas could be expanded or explored further.

We'll build two rockets to illustrate two ways to use Bluetooth low energy with iOS. The first rocket, cleverly called ST-1 for "SensorTag—first rocket," carries just the SensorTag itself. The SensorTag is pretty light, so this rocket can be small and light, too.

ST-2 is a larger rocket that carries both the iPhone and the SensorTag, as shown in Figure 7-1. This gets around a limitation of the ST-1, which will fly out of range of the iPhone. The ST-1 collects data for about the first 40 meters of the flight, after which the radio signal from the SensorTag gets too weak for the iPhone to maintain a connection. By carrying the iPhone right next to the SensorTag, the ST-2 is able to collect data for the entire flight. The disadvantages are that the iPhone is put at some risk, and the payload and rocket are much larger and heavier.

This chapter also walks through two programs, one to collect data and one to display and analyze the data.

After exploring how to do it all yourself, the chapter ends with a look at the data collected from several flights made by the author. Skip to "The Data" on page 189 to see how I analyzed the data I collected on my flights if you don't want to go through the steps needed to build your own rocket.

Parts Lists

Let's take a look at what you'll need to build the ST-1 and ST-2, and the additional items you'll want to have on hand.

ST-1

The ST-1 carries just the SensorTag, with remote sensing by a ground-based iPad or iPhone. Table 7-1 details what you'll need for the ST-1, shown in Figure 7-2.

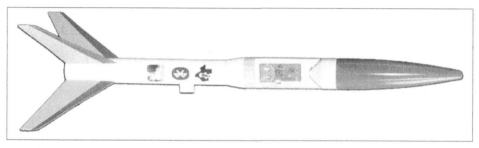

Figure 7-2. The ST-1

Table 7-1. Parts list for the ST-1

Part	Description
Estes Loadstar II Kit	Clear payload bays are hard to come by, and this kit has both the payload bay and all of the other parts needed for the rocket.
$\frac{1}{4}$" balsa sheet, about $1\frac{1}{2}$" x 4"	This will be used for the SensorTag holder.
$\frac{1}{32}$" x $\frac{1}{8}$" balsa strips, about 6" total	These strips are used to form grooves for the SensorTag holder.
$\frac{1}{16}$" x $\frac{1}{8}$" balsa strip, $1\frac{1}{2}$" long	This piece is used to offset the SensorTag from the bottom bulkhead.
B6-4 engine	Other engines can be used, but a less powerful engine is not recommended for this kit, and a more powerful engine is just going to carry the SensorTag farther out of range.
iPhone 4S or later, iPod 5th Gen or later, or iPad 3 or later	Used to record the data from the flight.

ST-2

The ST-2 rocket (shown in Figure 7-3) carries the SensorTag and an iPhone. You'll need a bigger rocket to handle the extra payload. Table 7-2 lists the parts required.

Figure 7-3. The ST-2

Table 7-2. Parts list for the ST-2

Part	Description
Two BT-80)body tubes	Note that Estes part 303090 – BT-80 Body Tubes is actually two body tubes, so you only need to buy one pair.

Part	Description
Two NC-80b nose cones	Yes, two. One is the nose cone you see; the other is used as a nose block to join the booster to the payload bay. You could also use a tube coupler and bulkhead.
24" nylon parachute	Estes part 002261 – Pro Series II 24" Nylon Parachute, or similar. You can also use two 18" nylon parachutes, one for the booster and the other for the payload.
Shock cord	Estes part 302278 – Shock Cords & Mount Pack, or a stiff elastic cord from the fabric store.
Engine mount	Estes part 303159 – D and E Engine Mount Kit, or similar.
$\frac{1}{8}$" x 4" x 36" balsa sheet	For fins and payload bay construction.
Two $1\frac{1}{2}$" x $\frac{1}{4}$" x 9" balsa strips or two $1\frac{1}{2}$" x $\frac{1}{8}$" x 18" balsa strips and four $\frac{1}{8}$" x $\frac{9}{16}$" x 9" strips	These strips are used to form the sides of the iPhone holder. The sides can be formed from $\frac{1}{4}$" strips using a router to form a channel, or they can be formed by gluing $\frac{1}{8}$" strips if a router is not available.
$\frac{1}{32}$" x $\frac{3}{8}$" balsa strips	Used for the payload bay construction.
$\frac{3}{8}$" balsa sheet, about 3" x 6"	Used for the payload bay construction.
Launch lug	Estes part 302320 – Launch Lug Pack, or similar.
iPhone 4S or later or iPod 5th Gen or later	Used to collect and display the data. If you have a choice, use an iPad for display; the larger screen is nice.

Other Items for Both Rockets

Several other items are also required for this project. Table 7-3 lists the other things you'll need.

Table 7-3. Additional requirements

Part	Description
TI SensorTag	This is the same device used in Chapter 6. We'll use the accelerometer, barometer, and gyroscope to analyze the rocket flight.
Building supplies	Paint, wood glue, five-minute epoxy, sandpaper, and a hobby knife—all stuff you probably have in your closet already.
Clear ink jet labels	Used to add the decals. Skip them if you like. Avery 5165 or similar.
Flash Programmer	Free software from Texas Instruments to update the SensorTag software. Available at *http://focus.ti.com/docs/toolsw/folders/print/flash-programmer.html*.
8G Firmware	Free firmware update for the SensorTag that changes the range of the accelerometer from ±2G to ±8G. It also increases the sample rate for the barometer. Get it from the Texas Instruments website at *http://processors.wiki.ti.com/images/9/9c/SensorTag_accel_8G.zip*.
CC Debugger	The CC Debugger is used to load new firmware onto the SensorTag, as well as for connecting diagnostic software.
Windows computer	The Flash Programmer only runs on Windows. You only need to load the firmware once, but plan ahead so you have access to a Windows computer for the firmware update.

Why Use a SensorTag?

Before we get started, it's fair to ask a very obvious question. This chapter shows how to collect acceleration, rotation, and pressure data from the SensorTag. Other than pressure, though, the iPhone can already do this by itself. It has an accelerometer, and most models have a gyroscope. Why bother with the SensorTag, other than for the pressure data?

While the iPhone has an accelerometer, the iOS software limits it to ±2G. That's just not good enough for a model rocket. We'll see that the rockets pull between 5G and 8G during the boost phase of flight, and can easily top 8G when the parachute deploys. Like the iPhone, the SensorTag is also normally limited to ±2G, but that's set in firmware and easily changed. The friendly engineers at TI have whipped off a special version of the firmware that supports ±8G. This firmware build is available from the TI website (*http://processors.wiki.ti.com/images/9/9c/SensorTag_accel_8G.zip*).

The SensorTag also gives us an option not available with the iPhone itself: We don't have to fly the iPhone at all to get the data. For about the first 40 meters (yards) of flight, we can hold our precious iPhone safely in our hands and still get flight data. While we still need to fly the iPhone for higher flights with more powerful engines, a high school physics class can easily collect data through the maximum velocity stage of the flight without risking an iPhone.

Construction

There are two ways to get data on your iPhone, and we'll build rockets for both. Let's start with the larger rocket that flies both the iPhone and the TI SensorTag in the same payload. Printed plans (*http://www.byteworks.us/Byte_Works/Blog/Entries/2012/10/31_Collect_Data_from_an_iPhone_Rocket_Flight_files/Plans%20Download.zip*) are available from the Byte Works website (*http://www.byteworks.us*).

ST-2: The iPhone/SensorTag Rocket

The booster is a pretty standard model rocket (shown in Figure 7-4). Assembly directions for model rockets are covered in lots of places, so we won't go over them here.

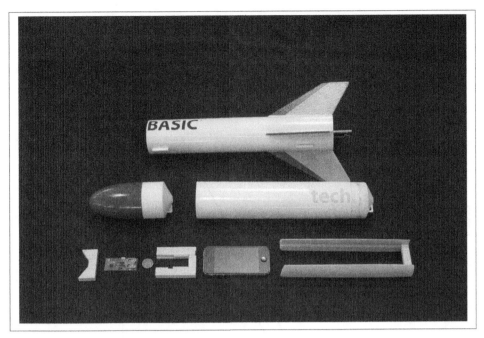

Figure 7-4. Exploded view of the ST-2

One thought, though, is to change the fin design. The fins shown here were chosen because the rocket was going to be shown at a trade show in Germany and needed to stand up on a table. The swept fins are pretty vulnerable to breakage on such a heavy rocket, though, and did break on two flights. A more practical design is also shown in the plans. This rocket won't stand up, but it is also less likely to break the fins when it lands.

Full-size fin patterns for both the design shown here and the alternate are shown in Figures 7-5 and 7-6. Be sure to match the fin grain when you cut the fins from 1/8 balsa sheets.

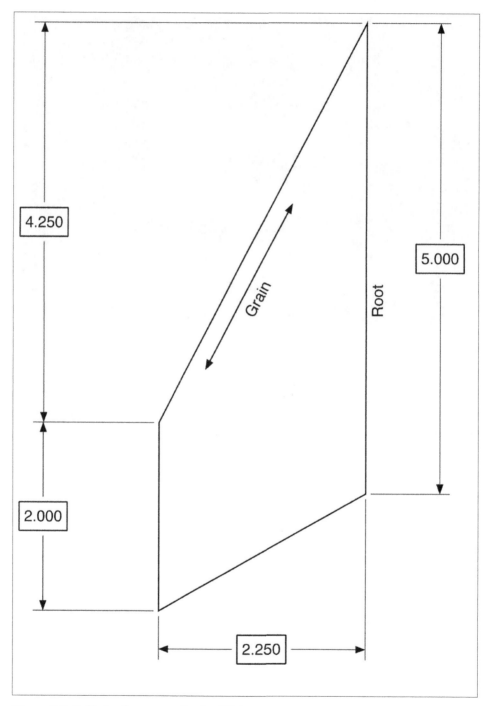

Figure 7-5. Full-size fin pattern for the ST-2

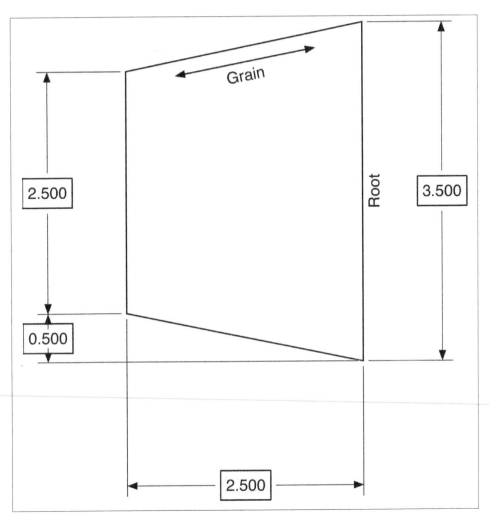

Figure 7-6. Full-size alternate fin pattern for the ST-2

The main part of the payload bay (see Figure 7-7) is also pretty standard. The only exception is the plug at the bottom of the payload bay. I could not find a nose plug for a BT-80 body tube, so I cut the top off of a nose cone and epoxied a $\frac{1}{8}''$ piece of balsa across the top to make my own. The finished plug was epoxied into the payload tube. I used epoxy because I wanted a very strong joint, but did not want to put too much of the nose cone in the payload tube—only about $\frac{1}{8}''$ of plastic was actually in the payload section.

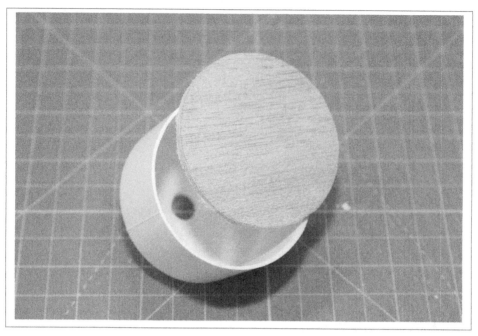

Figure 7-7. Construction of the payload bulkhead

Inside the payload section is a balsa wood holder for the iPhone and SensorTag (Figure 7-8). It's constructed using two side rails with a slot sized to fit the iPhone and other wood pieces, one of which holds the SensorTag. The dimensions shown in Figure 7-9 are for an iPhone 4S. If you have an iPhone 5, you will need to adjust the dimensions a bit to account for the thinner, taller model.

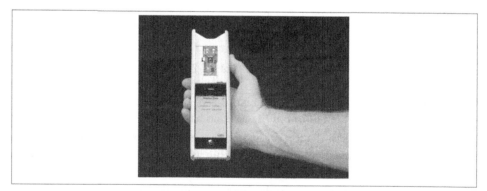

Figure 7-8. Completed iPhone and SensorTag holder

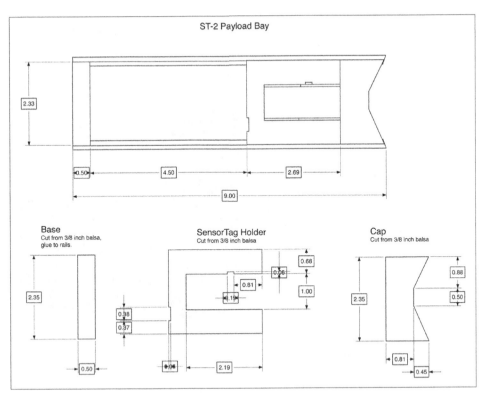

Figure 7-9. Plans for the ST-2 payload bay

There are two ways to build the rails. If you have a router, start with two pieces of balsa wood measuring $\frac{1}{4}'' \times 1\frac{1}{2}'' \times 9''$. Cut a $\frac{3}{8}''$ wide, $\frac{1}{8}''$ deep groove in the center of the strip. If you don't have a router, start with a $\frac{1}{8}'' \times 1\frac{1}{2}'' \times 9''$ piece of balsa and glue two $\frac{1}{8}'' \times \frac{9}{16}'' \times 9''$ strips onto it to form the $\frac{3}{8}''$-wide slot down the middle of the long piece.

With the pieces in place, the next step is to sand a curve on the outside of the pieces so they slide smoothly into the body tube (Figure 7-10). Place a sheet of sandpaper in a body tube and start sanding! When you are finished, the side pieces should lay snugly against the inside wall of the body tube. Glue a $2\frac{5}{16}'' \times \frac{3}{8}'' \times \frac{1}{2}''$ piece of balsa across the bottom of the side bars to form the main piece for the payload section.

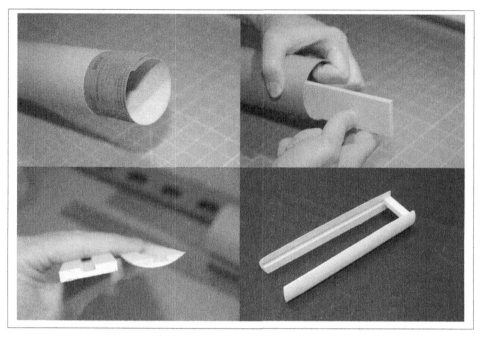

Figure 7-10. Construction of the side bars

The next section holds the TI SensorTag. Cut the pieces as shown in the plans. Since $\frac{1}{16}''$ router bits are tough to come by, this time use $\frac{1}{32}'' \times \frac{3}{16}''$ strips of balsa to form the groove for the SensorTag. Note that the groove stops about halfway up, leaving room for the tall components of the SensorTag. The notch in the side is positioned over the button on the SensorTag, leaving room to insert a fingernail and push the pairing button just before the launch. The notch at the bottom fits over the power button on the iPhone so the mount doesn't turn the phone on or off.

Cut the end piece from $\frac{3}{8}''$ balsa, making it just a tad longer than needed.

Don't glue these pieces into the side rails! Once complete, slide in the iPhone, slip the SensorTag into its holder, and slide the holder on top of the iPhone, and cap it off with the end piece.

With all of the pieces cut (Figure 7-11), assemble the payload section and trim the tops of the end cap and rails so they fit snugly against the bottom of the nose cone. Sand carefully to remove the last bits of extra balsa from the rails and end cap.

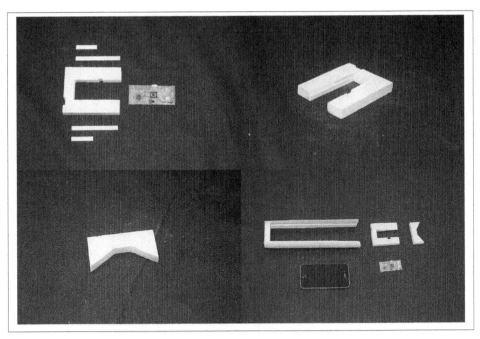

Figure 7-11. Construction of the SensorTag holder

ST-1: The SensorTag Rocket

If you're too chicken…uh, cautious to risk an iPhone in a rocket, there's the ST-1, which flies the SensorTag as a remote sensor (Figure 7-12).

I had good luck with a B6-4 engine, holding the iPhone in my hand while the rocket flew. Ground tests showed that the SensorTag signal is lost between 20 and 40 meters away, depending on how you hold the iPhone. I'd suggest trying this before the flight. With an iPhone 5, pointing either camera at the SensorTag gave about a 40–meter range. Pointing the top of the iPhone at the SensorTag dropped the range to 20 meters, while pointing the left, right, or bottom edge at it gave a range of about 30 meters. As expected, I lost some data during the flight, but still got enough to be interesting.

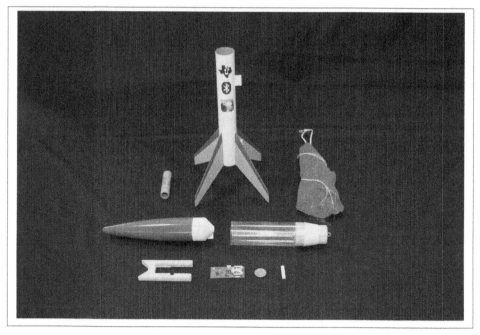

Figure 7-12. Exploded view of the ST-1

Construction is pretty straightforward. I started with an Estes Loadstar II Kit, making a couple of modifications. I added a fourth fin (Figure 7-13) to match the design of the larger ST-2, and substituted a longer, sleeker-looking nose cone from my parts bin. Neither change is really necessary; you can simply build the kit as designed and it will work just fine.

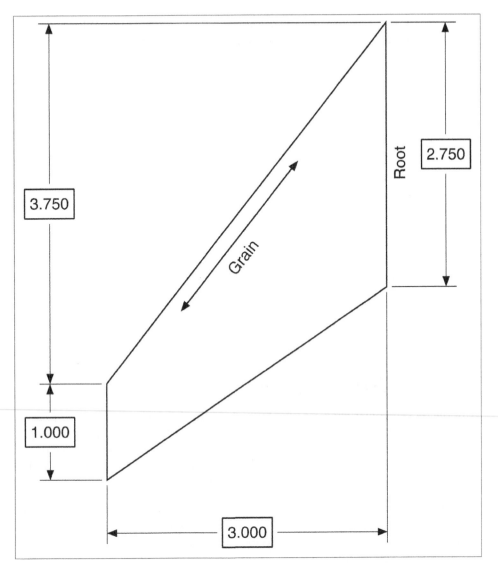

Figure 7-13. Fin pattern for the ST-1

The holder for the SensorTag is built much like the mount in the ST-2, as shown in Figure 7-14. The main form is cut from a single piece of $\frac{1}{4}''$-thick balsa. A notch is cut to make room for the pairing button. Cut strips of $\frac{1}{32}'' \times \frac{1}{8}''$ balsa and glue them to the inside edge to form a slot for the SensorTag. Once again, note that the strips stop about halfway down on one edge so the thicker components on the SensorTag near the battery have room. I added a $\frac{1}{4}'' \times \frac{1}{16}''$ strip at the bottom, carving a notch to accommodate it. This was really for aesthetics. You could easily omit this strip, making the cutout area

$\frac{1}{4}''$ shorter. For that matter, you could just stuff the SensorTag into the payload tube and pad it with a little cotton, but where's the fun in that?

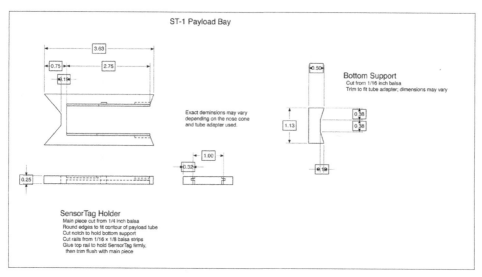

Figure 7-14. Plans for the ST-1 payload bay

The Data Collection Program

With the rockets constructed, it's time to work on the data collection program, shown in Figure 7-15. This program will have a very simple design, collecting information and saving it to a text file. The information will be written as it is collected, so even if there's a power loss during flight—or, more likely, at the time of landing (impact)—all of the data gathered up to the point of the power loss will be preserved.

The status is near the top of the screen. It uses a simple colored label to give a clear indication of whether the program is ready for the launch. The indicator starts off red, switches to yellow when the SensorTag is found, and finally changes to green once data is streaming in from all three sensors.

The next line shows the acceleration. This should vary a bit, especially as the SensorTag is jostled around. It should also read about 1. This gives a great check to make sure the correct firmware is installed on the SensorTag. If the SensorTag has the factory-installed firmware with the accelerometer range of ±2G, the value will be around 4.

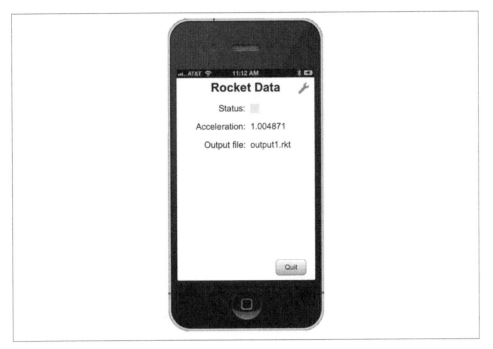

Figure 7-15. The Rocket Data program

Below this is a line showing the name of the output file. The program itself selects the filename, but showing the name gives you a chance to write down the correct filename for a specific rocket flight.

Finally, there is a Quit button in the bottom-right corner. Press the Quit button as soon as possible after the flight to stop data collection.

The first part of the program sets up global variables to hold the SensorTag peripheral object, the UUIDs for the services we will use, some handy constants for accessing the array of UUIDs, and the calibration data for the barometer:

```
! This program collects data from the TI Sensor Tag. It is designed
! for use in a model rocket, collecting acceleration, pressure, and
! rotation data. The data are written to a simple CSV text file for
! later analysis. Status information and the current acceleration
! are shown in real time.
!
! This program is designed to work with a special build of the firmware
! for the Sensor Tag that collects acceleration data in the range -8G
! to 8G, rather than the normal -2G to 2G.

! Set up variables to hold the peripheral and the characteristics
! for the battery and buzzer.
DIM sensorTag AS BLEPeripheral
```

```
! We will look for these services.
DIM servicesHeader AS STRING, services(3) AS STRING
servicesHeader = "-0451-4000-B000-000000000000"
services(1) = "F000AA10" & servicesHeader : ! Accelerometer
services(2) = "F000AA40" & servicesHeader : ! Pressure
services(3) = "F000AA50" & servicesHeader : ! Gyroscope
accel% = 1
press% = 2
gyro% = 3
services% = 0

DIM m_barCalib(8)
```

This program will have a simple GUI interface. After defining some controls that will be used in multiple subroutines, setUpGUI is called to create and display the user interface:

```
! Set up the user interface. Several globals are defined here and
! used by multiple subroutines.
DIM quit AS Button, status AS Label, accelValue AS Label
setUpGUI
```

Just as with the programs from Chapter 6, the next step is to start scanning for the SensorTag:

```
! Start the BLE service and begin scanning for devices.
debug = 0
BLE.startBLE
DIM uuid(0) AS STRING
BLE.startScan(uuid)
```

BLEDiscoveredPeripheral is called once iOS finds the SensorTag. It records the peripheral's object in the global variable sensorTag, begins the connection process, and stops scanning for peripherals:

```
! Called when a peripheral is found. If it is a Sensor Tag, we
! initiate a connection to it and stop scanning for peripherals.
!
! Parameters:
!    time - The time when the peripheral was discovered.
!    peripheral - The peripheral that was discovered.
!    services - List of services offered by the device.
!    advertisements - Advertisements (information provided by the
!        device without the need to read a service/characteristic)
!    rssi - Received Signal Strength Indicator
!
SUB BLEDiscoveredPeripheral (time AS DOUBLE, _
                            peripheral AS BLEPeripheral, _
                            services() AS STRING, _
                            advertisements(,) AS STRING, _
                            rssi)
IF peripheral.bleName = "TI BLE Sensor Tag" THEN
```

```
    sensorTag = peripheral
    BLE.connect(sensorTag)
    BLE.stopScan
  END IF
END SUB
```

BLEPeripheralInfo is called when the connection status changes. The first call is made right after the connection is established, with kind set to 1. The program sets the status color to yellow and begins discovery of the peripheral's services:

```
! Called to report information about the connection status of the
! peripheral or to report that services have been discovered.
!
! Parameters:
!    time - The time when the information was received.
!    peripheral - The peripheral.
!    kind - The kind of call. One of
!         1 - Connection completed
!         2 - Connection failed
!         3 - Connection lost
!         4 - Services discovered
!    message - For errors, a human-readable error message.
!    err - If there was an error, the Apple error number. If there
!         was no error, this value is 0.
!
SUB BLEPeripheralInfo (time AS DOUBLE, _
                        peripheral AS BLEPeripheral, _
                        kind AS INTEGER, _
                        message AS STRING, _
                        err AS LONG)
  IF kind = 1 THEN
    ! The connection was established. Look for available services.
    IF debug THEN PRINT "Connection made."
    status.setBackgroundColor(1, 1, 0): ! Connection made: Status Yellow.
    peripheral.discoverServices(uuid)
```

BLEPeripheralInfo is called with kind set to 2 or 3 if the connection is lost. Unlike the simple programs from the last chapter, this program takes aggressive action if the connection is lost. The status is colored red, then the program attempts to reconnect to the SensorTag:

```
  ELSE IF kind = 2 OR kind = 3 THEN
    IF debug THEN PRINT "Connection lost: "; kind
    status.setBackgroundColor(1, 0, 0): ! Connection lost: Status Red.
    BLE.connect(sensorTag)
```

Finally, BLEPeripheralInfo is called as the services are discovered, this time with kind set to 4. The program begins discovery of the characteristics for the three services—acceleration, rotation, and pressure—that it will record:

```
  ELSE IF kind = 4 THEN
    ! Services were found. If it is one of the ones we are interested
```

```
      ! in, begin discovery of its characteristics.
      DIM availableServices(1) AS BLEService
      availableServices = peripheral.services
      FOR s = 1 to UBOUND(services, 1)
        FOR a = 1 TO UBOUND(availableServices, 1)
          IF services(s) = availableServices(a).uuid THEN
            IF debug THEN PRINT "Discovering characteristics for "; services(s)
            peripheral.discoverCharacteristics(uuid, availableServices(a))
          END IF
        NEXT
      NEXT
    END IF
  END SUB
```

BLEServiceInfo is called as the characteristics for the services are discovered. The program uses three characteristics for the accelerometer. These were covered in Chapter 6, so the code should look familiar. The program starts the accelerometer, sets the data rate to 10 samples per second, and notes that the accelerometer has started. If the gyroscope and barometer are already running, the status indicator is set to green:

```
! Called to report information about a characteristic or included
! services for a service. If it is one we are interested in, start
! handling it.
!
! Parameters:
!    time - The time when the information was received.
!    peripheral - The peripheral.
!    service - The service whose characteristic or included
!        service was found.
!    kind - The kind of call. One of
!        1 - Characteristics found
!        2 - Included services found
!    message - For errors, a human-readable error message.
!    err - If there was an error, the Apple error number. If there
!        was no error, this value is 0.
!
SUB BLEServiceInfo (time AS DOUBLE, _
                    peripheral AS BLEPeripheral, _
                    service AS BLEService, _
                    kind AS INTEGER, _
                    message AS STRING, _
                    err AS LONG)
  IF kind = 1 THEN
    ! Get the characteristics.
    DIM characteristics(1) AS BLECharacteristic
    characteristics = service.characteristics
    FOR i = 1 TO UBOUND(characteristics, 1)
      IF service.uuid = services(accel%) THEN
        ! Found the accelerometer.
        SELECT CASE characteristics(i).uuid
          CASE "F000AA11" & servicesHeader
            ! Tell the accelerometer to begin sending data.
```

```
      IF debug THEN PRINT "Start accelerometer."
      DIM value(2) as INTEGER
      value = [0, 1]
      peripheral.writeCharacteristic(characteristics(i), value, 0)
      peripheral.setNotify(characteristics(i), 1)

  CASE "F000AA12" & servicesHeader
    ! Turn the accelerometer sensor on.
    IF debug THEN PRINT "Accelerometer on."
    DIM value(1) as INTEGER
    value(1) = 1
    peripheral.writeCharacteristic(characteristics(i), value, 1)
    services% = services% BITOR (1 << (accel% - 1))
    IF services% = 7 THEN
      ! Connection complete: Status Green.
      status.setBackgroundColor(0, 1, 0)
    END IF

  CASE "F000AA13" & servicesHeader
    ! Set the sample rate to 100ms.
    DIM value(1) as INTEGER
    value(1) = 10
    IF debug THEN PRINT "Setting accelerometer sample rate to "; value(1)
    peripheral.writeCharacteristic(characteristics(i), value, 1)
END SELECT
```

Starting the barometer and gyroscope follow a similar pattern. Refer back to Chapter 6 if any of the code seems confusing:

```
ELSE IF service.uuid = services(press%) THEN
  ! Found the pressure sensor.
  SELECT CASE characteristics(i).uuid
    CASE "F000AA41" & servicesHeader
      ! Tell the pressure sensor to begin sending data.
      IF debug THEN PRINT "Start pressure sensor."
      DIM value(2) as INTEGER
      value = [0, 1]
      peripheral.writeCharacteristic(characteristics(i), value, 0)
      peripheral.setNotify(characteristics(i), 1)

    CASE "F000AA42" & servicesHeader
      ! Turn the pressure sensor on.
      IF debug THEN PRINT "Pressure on."
      DIM value(1) as INTEGER
      value(1) = 1
      peripheral.writeCharacteristic(characteristics(i), value, 1)
      value(1) = 2
      peripheral.writeCharacteristic(characteristics(i), value, 1)
      services% = services% BITOR (1 << (press% - 1))
      IF services% = 7 THEN
        ! Connection complete: Status Green.
        status.setBackgroundColor(0, 1, 0)
      END IF
```

```
        CASE "F000AA43" & servicesHeader
          ! Get the calibration data.
          peripheral.readCharacteristic(characteristics(i))
      END SELECT
    ELSE IF service.uuid = services(gyro%) THEN
      ! Found the gyroscope.
      SELECT CASE characteristics(i).uuid
        CASE "F000AA51" & servicesHeader
          ! Tell the gyroscope to begin sending data.
          IF debug THEN PRINT "Start gyroscope."
          DIM value(2) as INTEGER
          value = [0, 1]
          peripheral.writeCharacteristic(characteristics(i), value, 0)
          peripheral.setNotify(characteristics(i), 1)

        CASE "F000AA52" & servicesHeader
          ! Turn the gyroscope on.
          IF debug THEN PRINT "Gyroscope on."
          DIM value(1) as INTEGER
          value(1) = 7
          peripheral.writeCharacteristic(characteristics(i), value, 1)

          services% = services% BITOR (1 << (gyro% - 1))
          IF services% = 7 THEN
            ! Connection complete: Status Green.
            status.setBackgroundColor(0, 1, 0)
          END IF

      END SELECT
    END IF
  NEXT
END IF
END SUB
```

As the sensors start to collect data, the operating system makes calls to `BLECharacter isticInfo` to let the program know the data is available. The data is collected just like it was collected in Chapter 6; the difference is what the program does with the data. Rather than printing the information to the console, the Rocket Data program writes it to a file. The file is opened by the `setUpGUI` subroutine, which we'll look at in a moment. The total G force is calculated by finding the length of the overall acceleration vector, and then placed in a text field. This is where the acceleration displayed on the apps screen is updated:

```
! Called to return information from a characteristic.
!
! Parameters:
!     time - The time when the information was received.
!     peripheral - The peripheral.
!     characteristic - The characteristic whose information
!         changed.
```

```
!   kind - The kind of call. One of
!       1 - Called after a discoverDescriptors call.
!       2 - Called after a readCharacteristics call.
!       3 - Called to report status after a writeCharacteristics
!           call.
!   message - For errors, a human-readable error message.
!   err - If there was an error, the Apple error number. If there
!       was no error, this value is 0.
!
SUB BLECharacteristicInfo (time AS DOUBLE, _
                          peripheral AS BLEPeripheral, _
                          characteristic AS BLECharacteristic, _
                          kind AS INTEGER, _
                          message AS STRING, _
                          err AS LONG)
IF kind = 2 THEN
  DIM value(1) AS INTEGER
  value = characteristic.value
  SELECT CASE characteristic.uuid
    CASE "F000AA11" & servicesHeader
      ! Get the acceleration.
      c = 16
      p% = value(1)
      IF p% BITAND $0080 THEN p% = p% BITOR $FF00
      x = p%/c

      p% = value(2)
      IF p% BITAND $0080 THEN p% = p% BITOR $FF00
      y = p%/c

      p% = value(3)
      IF p% BITAND $0080 THEN p% = p% BITOR $FF00
      z = p%/c

      PRINT #1, "acceleration,"; time; ","; x; ","; y; ","; z
      g = sqr(x*x + y*y + z*z)
      accelValue.setText(STR(g))
```

There is one key difference between the way the acceleration information is handled here and the way it was handled in Chapter 6. This subroutine divides the value reported by the sensor by 16 rather than 64, since the program is expecting the acceleration to range from –8G to 8G, not –2G to 2G. The firmware on the SensorTag needs to change to accommodate the higher acceleration range. We will look at updating the firmware right after we finish with this program.

```
    CASE "F000AA41" & servicesHeader
      ! Get the pressure.
      Tr = value(1) BITOR (value(2) << 8)
      S = m_barCalib(3) + Tr*(m_barCalib(4)/2^17 + Tr*m_barCalib(5)/2^34)
      O = m_barCalib(6)*2^14 + Tr*(m_barCalib(7)/8.0 + Tr*m_barCalib(8)/2^19)
      Pr = (value(3) BITOR (value(4) << 8)) BITAND $00FFFF
      Pa = (S*Pr + O)/2^14
```

```
      ! Convert from Pascal to Bar and use a display range of 0.6 to 1.2 Bar.
      Pa = Pa/100000.0
      PRINT #1, "pressure,"; time; ","; Pa

    CASE "F000AA43" & servicesHeader
      ! Get the pressure calibration data.
      FOR i = 1 TO 4
        j = 1 + (i - 1)*2
        m_barCalib(i) = (value(j) BITOR (value(j + 1) << 8)) BITAND $00FFFF
      NEXT
      FOR i = 5 TO 8
        j = 1 + (i - 1)*2
        m_barCalib(i) = value(j) BITOR (value(j + 1) << 8)
      NEXT

    CASE "F000AA51" & servicesHeader
      ! Update the gyroscope.
      c = 65536.0/500.0
      x = ((value(2) << 8) BITOR value(1))/c
      y = ((value(4) << 8) BITOR value(3))/c
      z = ((value(6) << 8) BITOR value(5))/c
      PRINT #1, "rotation,"; time; ","; x; ","; y; ","; z

  END SELECT
 END IF
END SUB
```

The barometer and gyroscope are handled the same way. We collect the values just like
we did in the programs from Chapter 6, then write them to the output file:

Next, the program generates a unique filename for the output file:

```
! Gets a unique filename for the output file. The filename begins
! with "output", followed by a number and then ".rkt".
!
! Returns: An unused filename.
!
FUNCTION getFileName AS STRING
index = 1
done = 0
WHILE NOT done
  name$ = "output" & STR(index) & ".rkt"
  IF EXISTS(name$) THEN
    index = index + 1
  ELSE
    done = 1
  END IF
WEND
getFileName = name$
END FUNCTION
```

`getFileName` is a utility function that finds the next available filename for an output file and returns it. The subroutine appends a number to the root filename, then checks to see if the file already exists. If it does, the next higher number is tried. The filename is returned as soon as one is found that doesn't exist already. This is a pretty simple way to get a unique name for a file. It will get a bit slow if there are hundreds of files, but that's not likely to happen for this application.

While the output file is ASCII text that can be opened by virtually any text editor or spreadsheet, a file extension of *.rkt* is used so the data analysis program can distinguish the files from other text files in the iOS sandbox. If needed, the file extension can safely be changed to *.txt* to make it easier to load into other programs.

The `setUpGUI` subroutine was called at the very start of the program. It sets up the user interface for the program and opens the output file. The first step is to change from the default mode of drawing the background using pixel graphics to drawing the background using vector graphics. We do that by turning pixel graphics off using the call `Graphics.setPixelGraphics(0)`. This gives a tiny performance boost when there is nothing to draw:

```
! Set up the GUI and the output data file.
!
SUB setUpGUI
! Use vector graphics.
Graphics.setPixelGraphics(0)
```

Next, we turn on full-screen graphics and only allow portrait orientation. We don't want the iPhone wasting valuable CPU time rotating the display as the rocket bounces around on a parachute!

```
! Switch to the graphics screen.
System.showGraphics(1)
IF System.device = 0 THEN
  System.setAllowedOrientations(1)
END IF
```

A label is used to place the program's title at the top of the screen:

```
! Get the size of the display.
height = Graphics.height
width = Graphics.width

! Label the app.
DIM title AS Label
title = Graphics.newLabel(10, 10, Graphics.width - 20, 25)
title.setText("Rocket Data")
title.setAlignment(2)
title.setFont("Arial", 28, 1)
```

Two labels are used for the status. The first shows the text "Status" while the second is just a colored square blob. The color is set to red. Only the first label is dimensioned here; the color blob was dimensioned as a global variable at the start of the program so other subroutines can access the label to change the color as the connection status changes:

```
! Add a status indicator. The status label is set to red initially,
! turns yellow when a connection is made to the Sensor Tag, and
! changes to green when all three sensors are started.
DIM statusLabel AS Label
y = 60
statusLabel = Graphics.newLabel(0, y, width/2)
statusLabel.setText("Status:")
statusLabel.setAlignment(3)
statusLabel.setFont("Arial", 20, 0)

status = Graphics.newLabel(width/2 + 10, y, 21)
status.setBackgroundColor(1, 0, 0)
```

The acceleration is shown with a pair of test labels. As with the status, the second label is dimensioned at the start of the program so it can be changed elsewhere:

```
! Add an overall acceleration indicator.
DIM accelLabel AS Label
y = y + 41
accelLabel = Graphics.newLabel(0, y, width/2)
accelLabel.setText("Acceleration:")
accelLabel.setAlignment(3)
accelLabel.setFont("Arial", 20, 0)

accelValue = Graphics.newLabel(width/2 + 10, y, width/2 - 10)
accelValue.setText("1")
accelValue.setFont("Arial", 20, 0)
```

Next the output file is opened, and the name of the output file displayed:

```
! Open the output file.
name$ = getFileName
OPEN name$ FOR OUTPUT AS #1

! Indicate the output filename.
DIM nameLabel AS Label, nameValue AS Label
y = y + 41
nameLabel = Graphics.newLabel(0, y, width/2)
nameLabel.setText("Output file:")
nameLabel.setAlignment(3)
nameLabel.setFont("Arial", 20, 0)

nameValue = Graphics.newLabel(width/2 + 10, y, width/2 - 10)
nameValue.setText(name$)
nameValue.setFont("Arial", 20, 0)
```

Suitably, the last step is to create a Quit button:

```
! Add a Quit button.
quit = Graphics.newButton(width - 92, height - 47)
```

```
quit.setTitle("Quit")
quit.setBackgroundColor(1, 1, 1)
quit.setGradientColor(0.6, 0.6, 0.6)
END SUB
```

`touchUpInside` is called when the user taps on a button. The only button on this program is the Quit button; if that button is tapped, the program makes sure any datafile is closed and stops:

```
! Handle a tap on a button.
!
! Parameters:
!    ctrl - The button that was tapped.
!    time - The time stamp when the button was tapped.
!
SUB touchUpInside (ctrl AS Button, time AS DOUBLE)
IF ctrl = quit THEN
  CLOSE #1
  STOP
END IF
END SUB
```

SensorTag 8G Software

The SensorTag comes with software already on the device for the normal sensor functions. The accelerometer on the SensorTag can be set to three ranges, ±2G, ±4G, or ±8G. The default firmware sets the acceleration to a range of ±2G, but this isn't high enough for a model rocket. While model rockets can exceed even 8G, the two rockets described here with the selected engines won't, so ±8G is ideal for our purpose.

The friendly folks at Texas Instruments took a special interest in this rocket project. As a result, they were kind enough to create a special version of their firmware that sets the acceleration range to ±8G. You can get the 8G firmware from the Texas Instruments website (*http://processors.wiki.ti.com/images/9/9c/SensorTag_accel_8G.zip*).

The next step is to install the firmware on the SensorTag. You will need a software package called Flash Programmer (*http://focus.ti.com/docs/toolsw/folders/print/flash-programmer.html*), which is also available on the Texas Instruments website. This software only runs on Windows, so you will also need a Windows computer for this step. Finally, you will need the CC Debugger, which makes the physical connection between the computer and SensorTag. This is available from Texas Instruments, and can be ordered along with the SensorTag itself.

The SensorTag needs power and doesn't get it from the computer, so start by inserting the battery. Use the small eight-conductor ribbon connector to connect the SensorTag to the small red circuit board. Be sure to connect it as shown in Figure 7-16, and center the connector on the pins. It's easy to get the connector on backward or shifted over one row when connecting it to the SensorTag, so double-check this step. Plug the red circuit board into the CC Debugger, and connect the CC Debugger to your computer using the USB cable. If everything is connected properly, the LED on the CC Debugger will glow green. If it's not green, make sure you remembered to install the battery in the SensorTag and check all of the connections carefully, especially where the ribbon cable connects to the SensorTag. Press the Reset button on the CC Debugger after making any changes. If the LED is not green, there is still something else connected incorrectly.

Figure 7-16. Installing firmware

After installation, run the SmartRF Flash Programmer (Figure 7-17) and set up the options exactly as shown in the screen capture. The path to the 8G firmware will be different on your computer, of course, but be sure you are pointed at the 8G firmware. The System-on-Chip section will be filled in automatically if you have correctly connected the SensorTag. With everything set up, click the "Perform actions" button to install the software on the SensorTag.

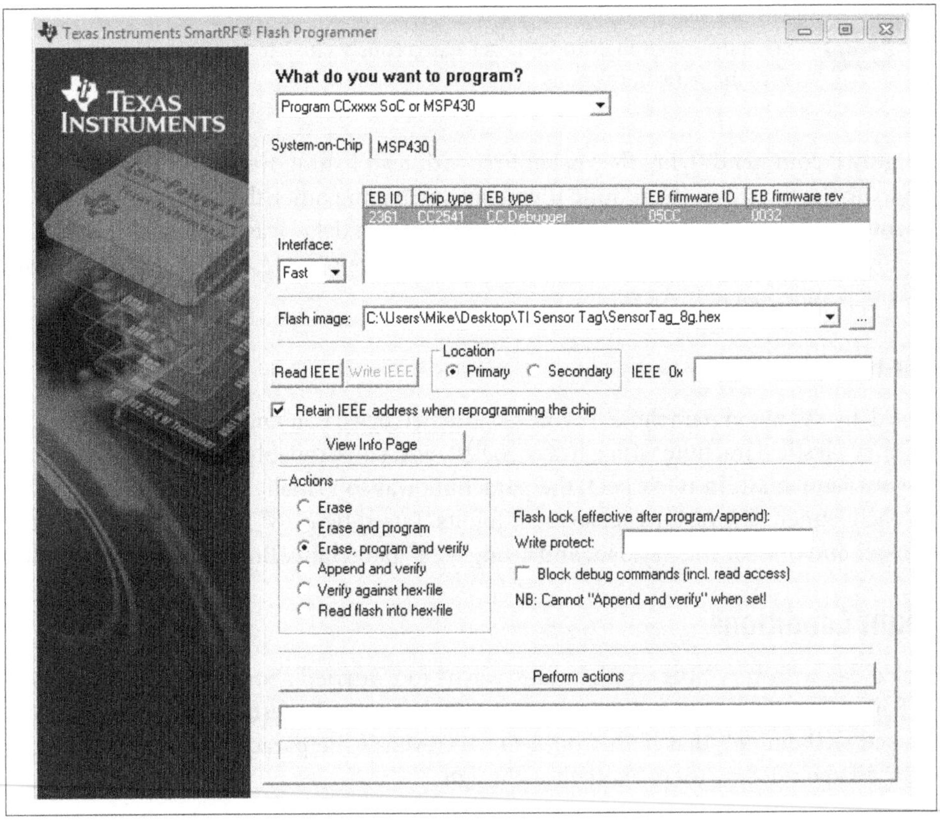

Figure 7-17. SmartRF Flash Programmer

Flight Tips

I'll share a few tips in this section. I did a few things right and a couple of things wrong, so you get the benefit of my mistakes!

Engines

I started with small engines. The B6-4 was perfect for the ST-1. You could, of course, use a C6-5, but expect to lose the signal for most of the flight if you do. Still, you will have all the data you need to calculate the maximum velocity, which is a pretty cool calculation.

I used a D12-3 for the ST-2 on the initial flights, and an E9-4 for subsequent flights.

Do not use the D12-5—this is a heavy rocket, and those extra two seconds are enough to let it get dangerously close to the ground before the parachute opens.

I highly recommend flying the rocket first with lead fishing weights in the payload to simulate the weight of the iPhone if you try any engine other than the D12-3 or E9-4. It would be sad to lose the rocket because the ejection delay was too long, but it would be a lot sadder to lose the SensorTag and iPhone. Even NASA does test flights before risking people in a new rocket.

Parachutes

I used an 18″ nylon parachute. If you've launched eggs, you know it's a bit risky to use a flimsy plastic parachute with a heavy rocket, and you probably have a few nylon ones in your launch kit. In retrospect, the parachute was too small, leading to a broken fin on each launch. On various subsequent flights, I used twin 18″ parachutes, one for the booster and one for the payload, and a single 24″ parachute. Both options work well.

Flight Conditions

Pick a day with very little wind. I really wasn't too worried about breaking the iPhone or SensorTag. While I'm not going to put this theory to the test, I suspect that they are packed well enough that both would survive even if the parachute failed. I was very worried about losing them altogether, though.

Some of my rocket flights are captured in a short video on YouTube (*http://www.youtube.com/watch?v=8YNjwcNXOK4*). As you can see from the movie, we have a great launch site here in Albuquerque, with almost no obstructions for hundreds of yards. I also picked a time with so little wind that, on one launch, the rocket landed closer to where I was standing than the launch pad.

Power Up!

Just before launch, start the Rocket Data app (Figure 7-15). The status indicator will start off red. Press the pairing button on the SensorTag and wait a few seconds. The status will change to yellow when the program finds the SensorTag, and to green once the accelerometer, gyroscope, and barometer start sending data to the iPhone. The current acceleration is shown, and it should update fairly often unless the SensorTag is perfectly still. Press the pairing button again if you don't see the status change to yellow and green after a few seconds.

Once you have a green light on the data program, finish any remaining preparations and then launch quickly. Turn the data recording off as soon as you can after the launch. Less extraneous data means less processing time and effort when analyzing the data.

The Data

The data collection program is very simple, dumping the information to a CSV datafile. The name of the file is shown on the app's main screen as the data is collected. You can move this file to a desktop computer for further analysis or, as you'll see in a moment, analyze it right on your iPhone or iPad. The techBASIC Quick Start Guides (*http://www.byteworks.us/Byte_Works/Documentation.html*) will walk you through moving the datafiles to and from techBASIC.

Analyzing the Data

Data analysis can be very simple or very elaborate. Let's start simple.

The datafile is a series of lines with several comma-separated values on each line. Here are a few lines from one of my rocket flights:

```
acceleration,372263872.779401004314,-0.09375,-0.9375,-0.015625
acceleration,372263872.7796189785,-0.09375,-0.9375,-0.015625
acceleration,372263872.808314025402,-0.078125,-0.890625,-0.015625
pressure,372263872.958292007446,0.837329
rotation,372263873.168398022652,7.606506,3.944397,-6.614685
acceleration,372263873.708128988743,-0.078125,-0.96875,-0.046875
acceleration,372263873.708782017231,-0.078125,-0.96875,-0.046875
```

Each line starts with a tag specifying the sensor that reported the data. Next is a time-stamp indicating when the data was collected. For acceleration and rotation, the next three values are the acceleration in Gs along the x-, y-, and z-axes, or the rotation in degrees per second along each axis. For pressure, the single value is the pressure in bars. And no, the barometer is not broken—we're a mile high here in Albuquerque, and the pressure is a bit lower here.

CSV files like this one are very easy to import into spreadsheets and databases, and also very easy to read into BASIC programs. Here's a short techBASIC program that reads the data from a rocket flight and plots the acceleration. It's preloaded in techBASIC in the *O'Reilly Books* folder; the program is called Rocket Acceleration:

```
! Scan the file and count the number of data points.
accelCount% = 0
startTime# = -1
name$ = "flight1i.rkt"
OPEN name$ FOR INPUT AS #1
WHILE NOT EOF(1)
  INPUT #1, tag$, time#, x, y, z
  IF tag$ = "acceleration" THEN
    IF startTime# = -1 THEN
      startTime# = time#
    END IF
    accelCount% = accelCount% + 1
  END IF
```

```
WEND
CLOSE #1
IF accelCount% > 16383 THEN accelCount% = 16383

! Read the acceleration data.
DIM accel(accelCount%, 2)
OPEN name$ FOR INPUT AS #1
index% = 1
WHILE NOT EOF(1)
  INPUT #1, tag$, time#, x, y, z
  IF tag$ = "acceleration" THEN
    IF index% <= accelCount% THEN
      accel(index%, 1) = time# - startTime#
      accel(index%, 2) = SQR(x*x + y*y + z*z)
      index% = index% + 1
    END IF
  END IF
WEND
CLOSE #1

! Create the plot.
DIM p AS Plot, d AS PlotPoint
p = Graphics.newPlot
d = p.newPlot(accel)
System.showGraphics
```

Figure 7-18 shows the result of running the simple data analysis program on *flight1.rkt*, one of the datafiles from my own rocket flights. The plot was panned and zoomed using swipe and pinch gestures to show the interesting part of the actual rocket flight.

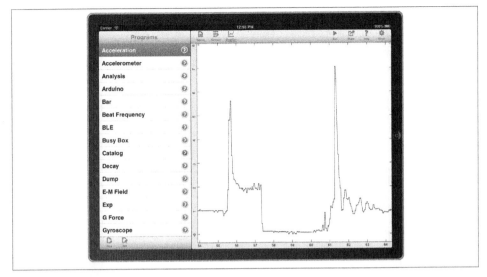

Figure 7-18. Data analysis of flight1.rkt

Rocket Data Analysis

While the previous short program shows how easy it is to analyze data, we're going to use a more sophisticated program (Figure 7-19) to look at the results from our rocket flights.

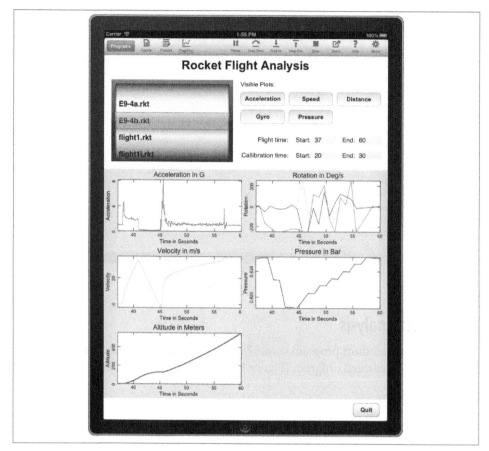

Figure 7-19. The Rocket Analysis app

This program is a bit more powerful and, as a result, a bit longer. It is included in both techBASIC and techBASIC Sampler; look for Rocket Analysis in the *O'Reilly Books* folder. While the program may be fun to explore and modify, it's not directly involved in data collection, so we won't go through it in detail in the book. Instead, we're going to use it to get a better look at what the sensor data from the rocket flights can tell us.

The picker lets you select from any available rocket data. The picker in Figure 7-20 is showing six datafiles from my rocket flights.

The buttons at the right of the picker let you show or hide the five plots used to analyze the data. Acceleration, shown in Figure 7-20, is the most interesting, so let's start there.

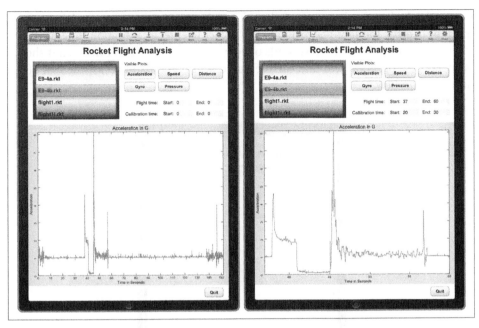

Figure 7-20. Acceleration

The original plot shows a lot of extra stuff, like the jostling as the payload is assembled, the long wait before the rocket launches, and the handling after the flight. We don't really need to see all of that. Set the start and end flight times by tapping the gray time box and entering appropriate values, and the interesting part of the data pops out. For this flight, I used a start time of 37 seconds and an end time of 60 seconds.

Now you can see the flight. As the rocket lifts off, the G force climbs rapidly to about 4.6G, then drops to about 2G and holds fairly steady for about two and a half seconds. The rocket then starts to coast, burning through a smoke charge that lets it move along at a bit more than 0G for a few more seconds. The ejection charge deploys the parachute, and the payload bounces around a bit during descent. An interesting point is that the parachute deployment, not the rocket engine, causes the highest G force.

Velocity and Altitude

But who cares about acceleration? We want to know how fast it went and how high it got, right? Not so fast—there are problems with that calculation. It seems so simple in physics class; you just multiply the acceleration by the time to get the velocity, and velocity times time to get distance. What could be easier?

Practice is a bit messier than theory. All measurements have error. There are three things that mess us up a bit when using an accelerometer to find speed and distance. The first is calibration error. The accelerometer is supposed to measure 1G sitting on the pad, but a quick look at the data shows it doesn't. It's close, but not quite there. We can limit this type of error, though. The program has a place to enter a calibration time, as shown in Figure 7-21. Pick a time 10 seconds or so before the rocket lifts off, when it should be at 1G. The program will find the measured acceleration and adjust the rest of the data appropriately.

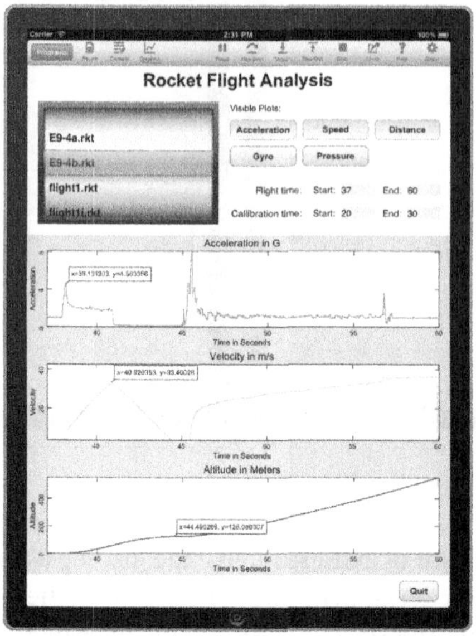

Figure 7-21. Acceleration, velocity, and altitude

So we're good, right? Not quite. Another kind of error occurs due to random fluctuations in the measurements. Let's see how this messes us up.

Let's say the first acceleration measurement is off by 0.02G, reporting 2.02G when it should have been 2.0G exactly. That's not much, right? After 0.1 seconds, we get:

$$v = 2.02 \cdot 0.1 = 0.202 \, m/s$$
$$d = 0.202 \cdot 0.1 = 0.0202 \, m$$

Not bad, we're only off by 0.02 mm. But now that error gets compounded on each subsequent calculation. Every single calculation will add another 0.0002 m to the distance that shouldn't be there. And that's just for a single, very small error on a single measurement—there are hundreds of chances to mess up, since there are hundreds of measurements.

The last, and most insidious, kind of error is systematic error. This is error that isn't random, but weights the results in the same direction each time a measurement is made. Calibration gets rid of a lot of the systematic error, but not all of it. Some is still there because of error in the calibration, and some creeps back in because we only calibrated for 1G. What happens if the accelerometer is fine at 1G, but off a bit at 2G, and off even more at 3G? This is just one way for systematic error to alter our results.

Finally, the program makes some simplifying assumptions. It assumes the rocket goes straight up and comes straight back down, but in reality, it will arc a bit. The program also assumes the rocket never experiences negative Gs, where drag is slowing it down at a faster rate than 1G. These assumptions add additional errors to the calculation. It's a great opportunity for further programming on your part. My gut feel is that the speed and altitude are probably good to plus or minus 20% or so up until the ejection charge fires, and essentially worthless after the rocket starts bouncing around on the parachute. In fact, the velocity and altitude information makes no sense after the rocket starts to bounce on the parachute. The calculations and observations of the height seem to bear that out. Proper error analysis is easily the topic of another complete book, though.

Blah, blah, blah. But how fast and high did it go?

Well, keeping in mind that the results could be off a bit due to error, this flight of the ST-2 with an E9-4 engine pulled 4.56G. The peak velocity was right after the engine burned out and the rocket started to coast, clocking in at 33.4 m/s or about 75 miles/hour (120 km/hour). The maximum altitude was 126 m, which is about 413 feet.

Rotation and Pressure

I wasn't expecting much from rotation or pressure measurements. Acceleration can be reported every 0.1 seconds, but pressure and rotation are only reported about once a second. These measurements could be pretty interesting on longer flights with bigger engines, though.

Rotation did tell me a bit (see Figure 7-22). The rotation was lower during powered flight than I expected, so I guess those fins were on straight! It looks like the rocket was making about a half turn a second. You can see rotation around y and z (green and blue, respectively) pick up a bit as the rocket arced over during the coast phase. Of course, once the parachute came out, the payload was rotating all over the place.

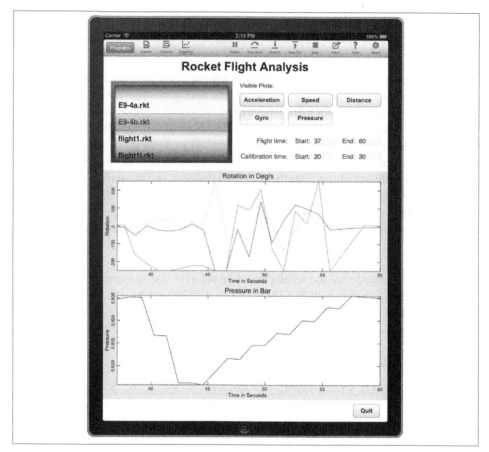

Figure 7-22. Rotation and pressure

Pressure shows a sharp drop as the rocket gains altitude, and a gradual rise as it parachutes back to earth. Remember the problems we discussed with using an accelerometer to find position? That's why high-power model rockets—the ones with H and larger engines—usually use a barometer, not an accelerometer, to find altitude.

What We Found

Let's take a closer look at the results from my sample data.

ST-1 Results

Two of the data sets, *flight1.rkt* and *flight2.rkt*, are from the ST-1, which only carried the SensorTag. As you can see, there is some data loss in each flight. The maximum acceleration on the first flight hit or exceeded 8G for a moment, then dropped to about 3G.

There was some data loss during the maximum boost phase on the second flight, but the data picked up again while the rocket was still pulling around 3G.

So why did a tiny B6-4 engine pull 8G while an E9-4 only got to 5.8G in my tests? The ST-1 rocket is much lighter, so from:

$$f = ma$$

we see that the lighter rocket can accelerate much faster. There are a lot of fun calculations here for a high school physics class. I won't spoil all of the fun, but I will pass on the launch weights of the rockets. The ST-1 weighed in at 114g, while the ST-2 was 396g with the D12-3 engine.

ST-2 Results

Two additional data sets, *flight1i.rkt* and *flight2i.rkt*, are from the ST-2 launched with D12-3 engines, and two more, *E9-4a.rkt* and *E9-4b.rkt*, are from flights with an E9-4 engine. The ST-2 carried the iPhone along with the SensorTag. One of the ST-2 flights was analyzed earlier; I'll leave you to explore the rest of the data I collected yourself or, hopefully, collect your own!

There are all sorts of changes and improvements you can make here, depending on your interests. Larger, more powerful rockets can carry the payload faster and higher. Different engines will yield different results. More careful data and error analysis can glean more from the information than shown here. The TI SensorTag also has sensors for humidity, temperature, and magnetic fields, and the iPhone has sensors for rotation and magnetic fields. There's a lot to explore as you build and adapt this project!

CHAPTER 8

Hacking a Radio-Controlled Truck with Bluetooth Low Energy and Arduino

About This Chapter

Prerequisites

Read Chapter 6 first to get a basic understanding of Bluetooth low energy technology.

Equipment

You will need an iPhone 4S or later or an iPod 5th Gen or later running iOS 5 or later, an Arduino Uno microcontroller, a RedBearLab BLE Shield, a hackable radio-controlled (RC) truck, and various small electrical parts to build this project. See Figure 8-1 for detailed parts lists.

Software

You will need a copy of techBASIC or techBASIC Sampler, a copy of the Arduino software for programming the Arduino microcontroller, and a copy of Firmata, an Arduino program that works well for our purposes. The Arduino software and Firmata are free downloads.

What You Will Learn

This project shows how to connect an iPhone or iPad to an Arduino microcontroller. The Arduino has limitless possibilities. Here you will see how to connect it to an H-Bridge to control motors, allowing an iPhone to steer a radio-controlled truck. This is a basic staple of robotics—all of the techniques are used in almost every mobile robot that uses DC motors for propulsion.

Controlling a Truck with BLE

This is a fairly challenging yet very rewarding project. There's something for everyone —cool control software for the discerning programmer, instructions on communicating with the Arduino microcontroller using Bluetooth low energy for the budding roboticist, a close look at using H-bridges to control motors for robotics, and some good old fashioned hardware hacking as we disassemble an inexpensive radio-controlled truck to make a few improvements (see Figure 8-1).

Figure 8-1. The BLE truck

We'll start by selecting and disassembling a radio-controlled truck that has a single speed for forward and reverse and can turn left or right, but offers no control over how much the truck turns. After a few measurements, we'll dispose of the radio that came with the truck and install an Arduino Uno microcontroller and a RedBearLab BLE Shield. This will allow us to control the truck from an iPhone or iPad using the same Bluetooth low energy commands seen in the previous two chapters.

Four digital outputs from the Arduino will be wired to a Texas Instruments SN754410 quadruple half-h driver. This chip provides two circuits that are a mainstay for robotics. The H bridge takes two digital inputs, forward and reverse, and supplies power to an electric motor to run it forward, run it in reverse, or turn it off entirely. We'll use one H bridge to control the steering and another to control the drive train. With the hardware built, the next step is to write the software. Rather than just providing forward and reverse, left and right like the original truck, we'll add full proportional control using a state engine to control pulse width modulation.

Finally, we'll hook all this up to the accelerometer in an iPhone (or iPad) so we can control the truck by tipping the iPhone itself.

Table 8-1 lists the parts we will use, along with some alternatives.

Table 8-1. BLE truck parts list

Part	Description
iPhone 4S or later, iPad 3 or later, or iPod Touch 5th Gen or later	We need an iOS device that supports Bluetooth low energy. If you are ambituous, you can skip to Chapter 12 to learn about the WiFly, replacing the RedBearLab BLE Shield with a WiFi connection, but you'll be working without a net at that point—it will work in theory, but there are no explicit instructions in this book.
techBASIC or techBASIC Sampler	We'll use the BLE Truck program, preinstalled in techBASIC.
RC truck	You have a lot of options here. The text discusses some things to consider when choosing a truck. I used the New Bright Radio Control Red Baja Extreme Ford F-150 at a 1:10 scale.
Arduino Uno microcontroller	The Arduino Uno is simply the most common and widely available model. If you are familiar with the differences, you could substitue some of the other models.
RedBearLab Bluetooth low energy Arduino shield	Available from Maker Shed, MKRBL1, or Seeed Studio, part SLD09041M.
Breadboard and jumpers	I used a small 17-row breadboard for the circuits. SparkFun part PRT-11658 or equivalent.
TI SN754410 H-Bridge motor Driver	SparkFun part COM-00315.
1x2 male headers (2)	These are a convenient way to connect the chip to the 1x2 Molex connectors you are likely to find in the radio-controlled truck. You can clip these from a longer strip, like SparkFun PRT-00116 Break Away Headers.
9V battery	Any standard 9V battery can be used to power the Arduino.
9V connector	SparkFun PRT-00091 or equivalent.
Mounting tape	Scotch double-sided mounting tape or equivalent.
SPDT or DPDT slide switch	Buy the truck first, and buy a switch that will fit. I used a switch from the Radio Shack 6-Piece Slide Switch Kit, catalog number 275-327.
Hookup wire	Seveal kinds of solid conductor 22 gauge wire are used. You can get by with less, but as shown, the project uses strands of colored wire, a few sections of four-conductor wire formed into a ribbon cable, and jumper wires for use on a breadboard.

Selecting a Truck

As you head off to the local Radio Shack or hobby store to look for a truck, there are a few things to keep in mind.

First, you're looking for an inexpensive truck. Sure, the fancy RC trucks look impressive. They are! But for this project, we need one of those cheap toys you usually avoid because they really aren't all that capable. If it costs more than $40, you're probably getting too fancy. I recently saw the model I used on eBay for $29.99. The truck should support left and right steering and forward and reverse motion. It does not need to have proportional control; you're going to create that yourself.

Pick a truck that is big enough to work with. The Arduino Uno and RedBearLab shield, with mounting tape and wires, will need a space about $2\frac{1}{2}'' \times 3\frac{1}{2}''$ in size with vertical clearance of $1\frac{1}{4}''$. The little red sports car may look cool, but a 15''-long monster truck will have more room.

Don't get one that is too big, though. Bigger vehicles require bigger motors, and you don't want to overpower the H bridge. If you're reasonably experienced with electronics, though, give in to temptation if you must. While this chapter does not have explicit directions, you can add heat sinks or daisy-chain the SN754410 to handle high-amperage motors, or you can scratch-build an H bridge using beefy power transistors. There's lots of help online, including a great tutorial by Chuck McManis (*http://www.mcmanis.com/chuck/robotics/tutorial/h-bridge/*).

Oh, and don't get the extended warranty. You're going to void it as soon as you get back to the workbench.

Disassembly

The disassembly instructions assume you are using the same New Bright Radio Control Red Baja Extreme Ford F-150 I used, but there are lots of similar models from the same company, and while the process for other trucks with similar capabilities will vary a bit, the basics will be the same.

Start by flipping the truck over. There are four screws that hold the body onto the chassis. Remove these screws as shown in Figure 8-2 and put them in a safe place—you'll need them later.

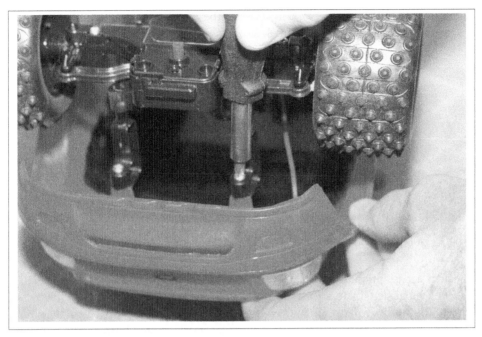

Figure 8-2. Remove the screws holding the body

Save yourself some time and make sure you only remove the screws holding the top onto the chassis. There are other screws there, too. Taking them off won't hurt anything, but some will take time to reassemble.

A single screw holds the plastic cover concealing the electronics for the RC controller. Remove the cover and discard it as shown in Figure 8-3.

Figure 8-3. Remove and discard the RC cover

There are two Molex connectors attached to the circuit board, one for each motor. Disconnect these as shown in Figure 8-4.

Figure 8-4. Remove the Molex connectors

The next step is optional if you're using exactly the same truck as me, but it's instructive. It's an essential step if you are using some other vehicle. We need to know what sort of current the motors will draw. That's essential for figuring out the capacity of the circuit we build. It's a great way to avoid melted, smoldering plastic and burned out electronics later.

Using some spare bits of wire, connect your multimeter into the circuit so one terminal of the Molex connector connects to the multimeter, and then connect the multimeter to the circuit board. The other terminal of the Molex connector connects directly back to its normal spot on the circuit board. Switch the multimeter to the amps setting; mine was labeled 10A, indicating the multimeter can handle 10 amps. Check where the leads are plugged into the multimeter—it's generally a different connector for measuring current. The connector on my multimeter was labeled 10ADC. The ground connection is the same.

With all of the preparation done, we're going to measure the current required to turn the wheels of the truck. As you can see in Figure 8-5, I got 0.75 amps—well within the operating limits for the SN754410, which can handle 1 amp per channel.

Figure 8-5. Checking the steering current

Now it's time to check the power needed to drive the truck. We'll check this twice, once with the rear wheels lifted off of the table, and once with the truck held so the wheels cannot turn. The current will be quite a bit different for these two measurements—the motor doesn't need much power to spin the wheels freely, but it will strain as hard as possible to move the truck when held still. The actual current in normal operation will be somewhere between these two values. As you can see from Figure 8-6, I got 0.18 amps when the wheels were off of the table, and 1.38 amps when the wheels were held still.

Figure 8-6. Checking the drive current

Now that's a concern. The SN754410 is rated at 1 amp per channel, but with the wheels locked, the motors can draw more. There are a few options here. One is to use a heat sink, which allows the chip to dump excess heat faster, increasing its ability to handle current. Another possibility is to use a different chip, or to use a pair of chips with each handling half of the current load. Still another possibility is to custom-build a large-capacity H bridge from large power transistors. Then there's the pragmatic approach, which is the one I used: plug it in and see how it works. After all, the normal operating current will be lower than the stalled current, and the chip is plugged into a breadboard, which already acts as a small heat sink.

I'll save you the suspense. With this truck, the SN754410 works just fine. I didn't detect any noticeable heating, even with the truck was stalled. Your results may vary, especially with a different vehicle, so check the chip to see if it is getting warm. If so, alter the design by adding a heat sink or using an H bridge that can handle more current.

What Happens if the Truck Is Too Big?

One of the book's reviewers pulled an old truck from a toy bin that was a bit larger than mine, pulling 4 amps, and gave the circuit a try. The truck did work for a while, but eventually the motor started to pulse. There may be a limiting circuit in the SN754410 that helps prevent damage, or perhaps he just got lucky.

While experimenting like this is a lot of fun, keep in mind that the circuits can burn out. The likely result is to destroy the SN754410. The only real hazard is the heat generated, which can be enough to melt plastic.

 Be vigilant for heat problems if you overextend the circuit.

The last step in disassembling the truck is to remove the radio. It's fastened to the vehicle in two ways. The first is a mechanical clip molded into the plastic body. Pop that loose.

The radio is also soldered to the battery connections from the battery compartment of the truck. Use a soldering iron with the heat set a bit higher than normal to soften the solder while pulling up gently near the connection to release the circuit board. You can also try a braided copper solder wick like the one shown in Figure 8-7, which absorbs the solder as it melts. Braided copper for use as a solder wick is available from Radio Shack and most electronics parts suppliers. It's a little neater than just loosening the solder, leaving clean holes to resolder the circuit board at a later date.

Figure 8-7. Removing the radio

Hacking the Truck

With the original radio control equipment out of the way, it's time to install the new circuitry. The heart of the circuit is the H bridge. That's a familiar circuit to most people who design and build small robots, but it may be new if that's not something you've done before, so let's start by getting a basic understanding of how the H bridge works.

The H Bridge

The fundamental problem we are trying to solve is how to turn a digital signal from the output pins on the Arduino microcontroller into forward, reverse, or zero current to power the drive motor and steering motor of the truck. Each of the motors can be in one of three states: no power, forward, or reverse. Since three states are possible, we'll need at least two bits of output from the Arduino to control a motor. If both bits are low (the 0 state), no power will go to the motor—it will just coast. If the first bit is high, we want to supply power to the motor to propel the truck forward or turn it left; bringing the other bit high should drive the truck backward or, in the case of the steering motor, turn it right. We won't use the state where both bits are high.

An H bridge is designed for exactly this purpose. Figure 8-8 shows a conceptual circuit for an H bridge. It also shows where the circuit gets its name. If you squint just right, concentrating on the wires connecting the switches and motor while ignoring the ones to the battery, you can see that the wires form the letter H.

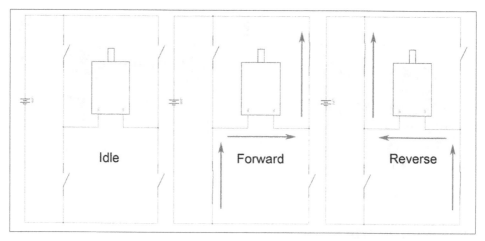

Figure 8-8. Conceptual circuit for an H bridge

The leftmost of the three circuits shows the situation when both bits are low. All four switches are open, so no current goes to the motor. Electric motors in this state not only don't move but don't resist turning much, either, so the truck can coast along if it was already moving.

The middle circuit shows the state when the first of our control bits is high. The lower-left and upper-right switches close, allowing current to flow through the motor, turning it in a direction we'll tap for forward motion.

The rightmost circuit shows what happens when the second control bit is high. The current flows, but it flows in the opposite direction, reversing the direction of the motor. In the case of the drive motor, the truck moves in the opposite direction. For the steering motor, the truck turns the opposite way.

In a real H bridge, the mechanical switches from Figure 8-8 are replaced with transistors or relays. Additional components are added to switch a single digital signal so it powers two different switches, preventing current from flowing the wrong way with odd switch configurations, and so forth. It's a very fun topic to explore. If you'd like to learn more, start with Chuck McManis's excellent introduction (*http://www.mcmanis.com/chuck/robotics/tutorial/h-bridge/*).

With this basic understanding of the H bridge, it's time to take a look at the Texas Instruments SN754410 chip.

The TI Chip

We'll discuss some of the circuit here to help you better understand the chip, but don't start wiring just yet. The complete circuit diagram appears in the next section; it won't make sense unless you understand how the chip is used first, though.

Orienting Yourself with an IC

Our other projects have used simple parts or entire circuit boards with pins that are labeled right on the device. The SN754410 is different. This circuit is packaged in what's called a *DIP*, which stands for dual inline package. There are other package types for chips, but this is a very common one for breadboard circuits since the pins are all 0.1 inches apart, just like the pins on the breadboard, and the distance between the pins is a multiple of 0.1 inches.

The pins on the chip are usually not labeled. In fact, the chip itself is often poorly labeled —you might have to hold it just right under a bright light and use a magnifying glass to read the chip information. Some may have a circular indentation, too, but there will almost always be a notch (see Figure 8-9) that will distinctly touch the edge of the chip. This is the top.

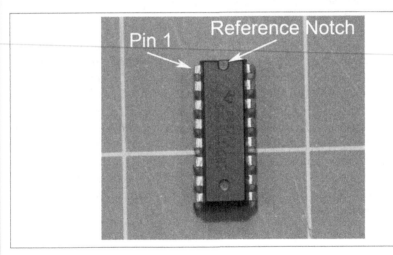

Figure 8-9. SN754410 notch location

Pins are numbered starting with 1, where pin 1 is immediately to the left of the notch, and proceeding in a counterclockwise direction when looking at the top of the chip, as shown in Figure 8-10. A few other pin numbers are included in the figure to keep you oriented.

This tiny little powerhouse, shown in Figure 8-10, actually contains two H bridges and two separate inputs for power, one for the chip itself and another for the motors.

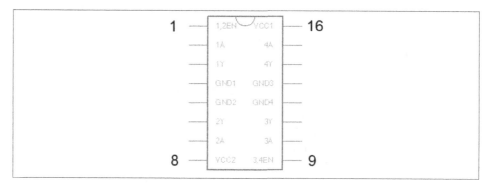

Figure 8-10. The SN754410

The four digital input pins are labeled with a number followed by an A. We'll wire pins 1A and 2A to control the drive motor, and 3A and 4A to control the steering motor. Those four pins will connect to digital output pins 2 through 5 on the Arduino.

The four power output pins are labeled with a number followed by a Y. Pins 1Y and 2Y will connect to the two connectors on the drive motor, while pins 3Y and 4Y will connect to the connectors on the steering motor.

Which pin is which? Well, if you wire them exactly as shown in the circuit diagram, set up the circuit exactly as shown in the photos that follow, and make sure the blue wires from the motors are to the outside and the brown wires to the inside, everything will work. And if you mess up? The truck will move backward when you tell it to move forward, or turn left when you tell it to turn right. Gosh, you'll have to turn one of the Molex connectors around. Frankly, it's easier to not worry about whether you are connecting the correct digital output and motor terminal to the correct pin. As long as pins 3 and 4 from the Arduino connect to the same side of the chip as the wires running to the motor, and Arduino pins 5 and 6 connect to the same side as the steering motor, the worst thing that will happen is the directions will be reversed, and you'll have to turn the connector around.

Don't screw the top of the truck back on until you check which direction the truck moves and turns when you try the software. Just be sure the A pins go to the Arduino and the Y pins go to the motors—applying the motor power to the Arduino would be bad.

Pins 1 and 9 are digital inputs that control whether each side of the circuit is active. We always want both sides to be active, so we'll wire them to +5V to keep them high at all times.

Pin 16 is the power input for the chip itself. We'll wire this to pins 1 and 9, and also to +5V on the Arduino, swiping digital power from the microcontroller so we don't need to build a separate power regulator.

Pin 8 is the power input for the motors. This gets wired to the positive terminal of the main battery compartment.

The central pins—4, 5, 12, and 13—are all ground pins. They also provide a heat sink for the chip. All four pins get wired together and connected to both the Arduino ground pin next to the 5V connector and the negative terminal on the main battery pack.

Wiring the Complete Circuit

We discussed much of the wiring when we looked closely at the SN754410 H bridge chip, but there are other details we need to consider. Figure 8-11 shows the complete wiring diagram. It's a fairly big, complicated diagram, so it's shown separately from the photos of the truck circuit, which are in Figure 8-12.

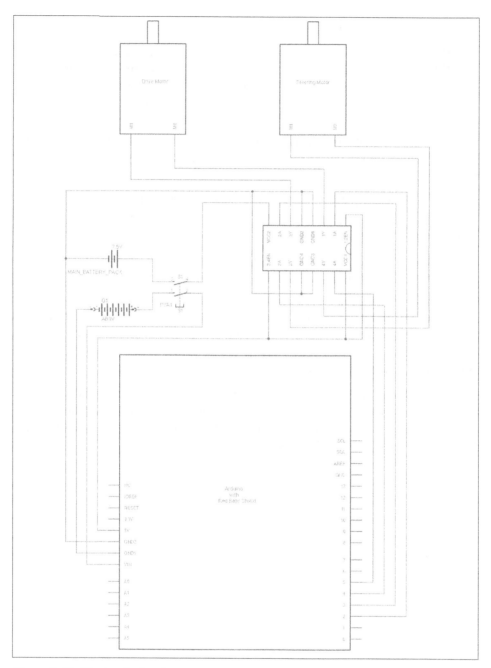

Figure 8-11. The truck circuit

Figure 8-12. The completed truck

The first step in wiring the circuits is to attach the various components to the truck.

Start with the switch (Figure 8-13). Remove the one that came with the truck and replace it with an SPDT or DPDT switch. A dab of super glue will hold it in place.

Figure 8-13. The switch

It's easier to connect wires to the switch and battery terminals before the remaining components are in place. Use 22-gauge solid-core wire for all wiring. It's convenient to have several colors, and even more convenient to have at least one section of ribbon cable with four colored wires joined together in a strip. You can get all of this wire at Radio Shack if you don't already have it in your parts box. Be creative with repurposing wire, though. There are several sections in a typical Radio Shack store for wire; I found the ribbon cable shown in the section marked for intercom wire.

Solder one wire—conventionally a black one—to the negative terminal on the 7.5V battery compartment. Solder a second wire to the positive terminal, and the other end to the central pole on the switch.

Solder three other red wires to the other poles of the switch. The wire leading to the central pole not connected to the 7.5V battery pack will eventually connect to the positive terminal of the 9V battery. The wire connecting to the corresponding pole will eventually plug into the RedBearLab board, and the last wire eventually connects to the breadboard.

Switch Designations

There are an enormous number of different kinds and sizes of switches (see Figure 8-14).

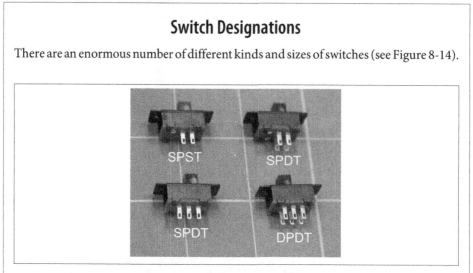

Figure 8-14. Switch types

When discussing the simpler switches, like slide switches, you'll usually see the designations SPST, SPDT, DPDT, and DPST. P and T stand for pole and throw. A pole is a connection for a switch, while throw is the number of connections the switch makes. The S and D stand for single and double. A common light switch is an SPST switch: it switches one thing on or off.

An SP switch has a single pole, so it is either on or off. A DP switch can route the current in two directions. For example, it can be set up to turn on one of two lights, alternating which of the lights is on as the switch is flipped.

An ST switch is used to turn one circuit on or off, while a DT switch controls two circuits with a single switch.

The switch in our truck needs to turn two circuits on or off. One is the power from the 9V battery used to power the Arduino, while the other is the 7.5V battery pack that comes with the truck. We just need to turn them on or off, so the ideal switch is an SPDT switch. It will have four connections on the bottom. A DPDT switch can also be used; in that case, we just ignore the extra two poles on one side of the switch, treating it like an SPDT switch. If this is new to you, hook a multimeter to some switches and make sure you understand how they work.

The easiest way to attach the Arduino to the truck is to use several layers of mounting tape, as shown in Figure 8-15. After removing the last red layer, press the Arduino onto the tape.

Figure 8-15. Mounting tape to mount the Arduino

The RedBearLab BLE Shield plugs directly into the Arduino. Figure 8-16 shows the Arduino Uno from the top and the RedBearLab BLE Shield from the bottom so you can see how the pins match up. The six pin connectors on the right in the figure are a great way to get everything lined up properly. Plug the two components together, remove the last strip of red plastic tape from the stack of mounting tape, and press the Arduino onto the sticky surface to secure these two components in place.

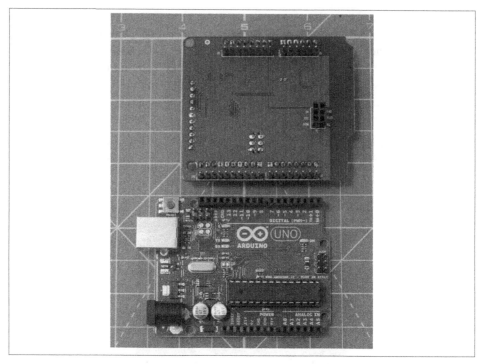

Figure 8-16. The RedBearLab BLE Shield (from the bottom) and Arduino Uno (from the top), showing how the pins align

It's easiest to wire the breadboard before inserting it into the truck. I used one with 17 rows of pins. Place the SN754410 with the notch to the right, leaving four rows of pins on the left and five on the right, as shown in Figure 8-17.

Figure 8-17. Mount the chip on the breadboard with the notch to the right

I used precut, color-coded jumper wires to make the connections on the breadboard shown in Figure 8-18. Don't miss the gray wire on the bottom row or the two uncolored jumpers that connect pins 4 and 5 and pins 12 and 13. Jumper wires that connect adjacent pins don't generally have colorful insulation, so they are easy to miss if you are not looking for them.

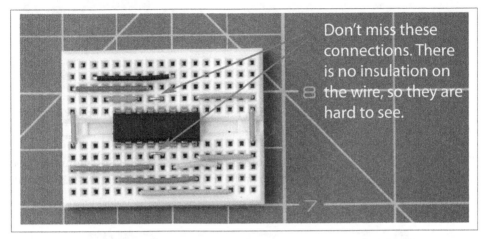

Figure 8-18. Use jumper wires or 22 gauge wire to wire the breadboard

Figure 8-19 shows the headers used to form connectors for the Molex connections from the truck motors. These generally come in long strips that can be cut into shorter lengths with a pair of diagonal cutters. The pins are not permanently fastened to the plastic, which is a good thing in our case. The pins on the left and the long set of pins on the top show how the pins are usually spaced. Push them so the pins are about halfway through, as seen on the bottom right, then mount the headers in the breadboard as shown.

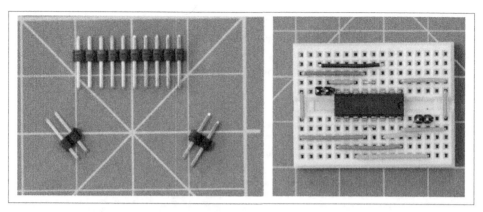

Figure 8-19. Cut two two-pin headers, push the pins about halfway through the plastic, and mount the headers on the breadboard

The breadboard may come with a layer of tape already attached. If not, mounting tape works fine. Attach it in the front section of the truck. Here it is in Figure 8-20 after all of the final wiring is done.

Figure 8-20. The breadboard

Attach the 9V battery using mounting tape—it fits well in the rear of the truck.

The wires from the rear motor won't quite reach the front circuit board. Cut the wires and splice extra wire to make them long enough, using shrink tubing or electrical tape so the bare wires won't touch. The leads from the 9V battery connector will also need to be extended.

You're almost there. Complete the wiring as shown in the preceding wiring diagram and photos, then double-check the connections. Don't forget the jumpers that connect the ground pins on the SN754410 on each side of the chip. They help with heat dissipation, but they're easy to miss in the photo since jumper wires to go between adjacent breadboard holes are bare—there is no colored insulator.

The last step is to reinstall the upper body, but don't do that until you've tested the truck in operation to make sure forward is forward and right is right.

Controlling the Arduino Uno

The Arduino is a microcontroller; it's a specialized computer in its own right. It may not have a keyboard or monitor (most of the time, anyway), but it definitely has an operating system. It also needs to be programmed.

What we need is a program that will allow us to send a command over the serial interface provided by the RedBearLab BLE Shield. The Arduino has a number of input and output pins. We need a program that will let us use these serial commands to turn Arduino digital I/O pins 2, 3, 4, and 5 on or off.

We could write this software ourselves, but there is already a nice open source software package that will do all of this and more. It's called Firmata. Our next step is to install the Firmata software on the Arduino so it knows how to respond to the serial commands that the iPhone software will send.

If you've used an Arduino before, much of what follows will be old hat for you. Skip or skim as appropriate.

Installing Arduino

Arduinos are programmed from a desktop computer. The software to program an Arduino is open source, and is cleverly called Arduino. This is the software we will use to move the Firmata firmware from the desktop computer to the Arduino. You can get a version for your computer from the Arduino website (*http://arduino.cc/en/main/software*). Versions are available for Linux, OS X, and Windows. Go to the website and download the Arduino software, install it, and run it.

The Arduino software is written in Java, so you also need to install a Java runtime. You may have already done that for another program. If not, you will get an error appropriate for your operating system. For OS X, you'll get a dialog stating that the software needs a Java runtime and asking if you would like to install one. Click Yes. On other platforms, you'll need to install Java from the Java website (*http://www.java.com/en/*). Follow the instructions for your particular platform to download and install Java.

Is Java Secure?

You may have heard about Java in the news recently, when many security experts recommended turning Java off due to security concerns. I just told you to install it. So, is your computer safe? Is Java secure?

Well, yes and no.

Java is used for two distinct reasons on your desktop computer, and a third on many servers. Java can run programs embedded in websites. These can have malware, and some security holes have been found in Java that caused the security concerns. These holes apply to web browsers, where Java programs can run automatically when you visit a web page. Earth is currently inhabited by 7 billion fairly nice people and a few real jerks. Because of the latter, you should leave Java turned off in your browser preferences unless you take the time to learn how to enable it safely.

Java programs like Arduino, though, are just desktop programs like any others. They don't have any special security holes, nor do they have any special protections. They are as safe as programs written in other languages.

So go ahead and install Java and use it for the Arduino software. Just don't enable it in your browser until you've learned enough to do it safely.

You'll see a screen like Figure 8-21 when you run the Arduino software.

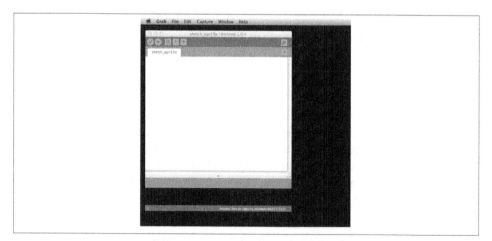

Figure 8-21. The Arduino software

Downloading Firmata

You can find general information about Firmata for Arduino at firmata.org (*http://firmata.org/wiki/Main_Page*). There is a link on the main page you'll want to explore: the Protocol link in the navigation section on the left gives a very terse description of the commands used to control an Arduino with Firmata. We'll go through all of the ones needed for the BLE Truck program in detail, but this is the document you will come back to if you want to explore other possible uses for Firmata.

There are a number of versions of the Firmata software. Rather than wade through all of them, go to the RedBearLab website (*http://redbearlab.com/bleshield/*), scroll down to the Quick Start section (see Figure 8-22), and click on the link BLE Shield Library for Arduino (*https://github.com/RedBearLab/Release/raw/master/BLEShield/BLEShield_Library_v1.0.2.zip*). This is the version of Firmata needed for full compatibility with the RedBearLab BLE Shield.

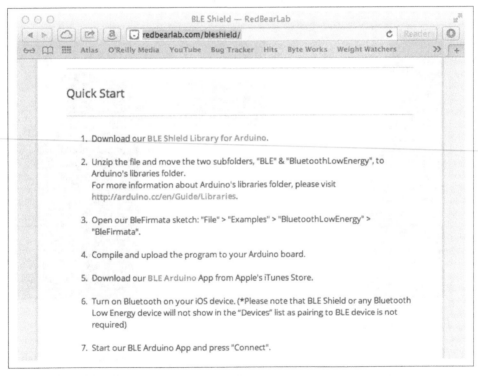

Figure 8-22. RedBearLab Quick Start section

Installing Firmata

The instructions on the RedBearLab website will lead you through installing Firmata in the Arduino program, but they don't cover moving the Firmata software to the Arduino hardware. Here are the steps required:

1. Unzip the file if your OS didn't do that automatically.

2. Locate the Arduino library folder. On Linux, it's located in the *libraries* folder of the Arduino sketchbook. On OS X, you'll find it at *Documents/Arduino/libraries*. On Windows, it's at *My Documents\Arduino\libraries*.

3. Copy the folders *BLE* and *BluetoothLowEnergy* from the unpacked ZIP file to the *library* folder located in step 2.

4. If you are running Arduino, shut down the program and restart it.

5. Open the BLEFirmata library (see Figure 8-23). It's in the submenu File→Sketch-Book→libraries→BluetoothLowEnergy→BLEFirmata.

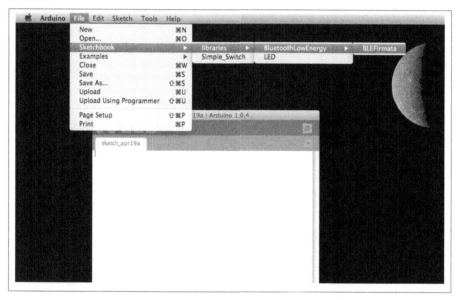

Figure 8-23. Open the BLEFirmata library

6. Connect the Arduino to your computer using a USB cable. It is safe to do this while the Arduino is installed in the truck, but you might as well turn the power off in the truck first.

7. Make sure your exact Arduino model is selected in the Tools→Board menu, as shown in Figure 8-24.

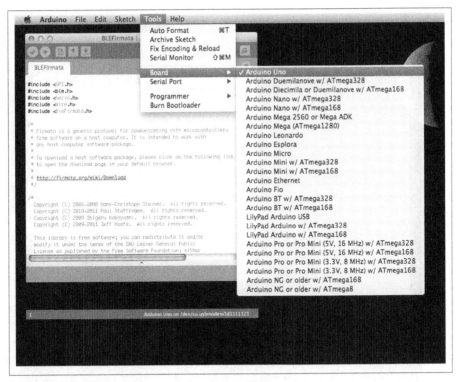

Figure 8-24. Make sure your Arduino model is selected

8. Check the Serial Port in the Tools→Serial Port menu. Make sure it corresponds to the Arduino. If you are not sure which serial port belongs to the Arduino, unplug it, check the menu, then plug it back in and check the menu again. Select the serial port that appears only when the Arduino is plugged in, picking the one that starts with *cu*, as shown in Figure 8-25.

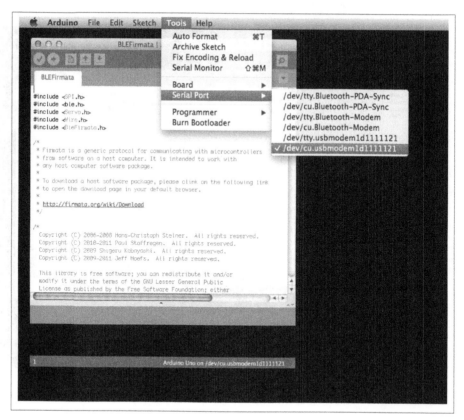

Figure 8-25. Select the proper serial port

9. Click the Upload button to install the Firmata library on the Arduino, as shown in Figure 8-26.

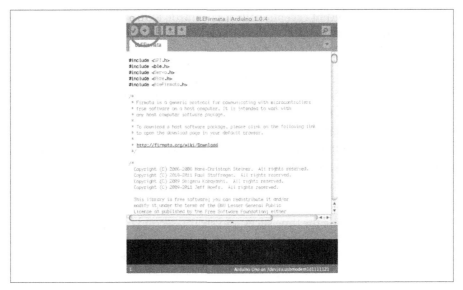

Figure 8-26. Click the Upload button

The Software

There are two important concepts embedded in the software. Before looking at the software itself, let's take a moment to look at these ideas.

Pulse Width Modulation

Back at the start of the chapter, I mentioned that we would improve on the capabilities of the original truck controls. The original radio could control speed and direction, but only using full power or none at all. We're going to implement proportional control for both. Proportional control means you can go slow or fast, or turn a little or a lot. The original truck could not do that. The circuit we've designed can't do it, either. It just turns the motors on or off.

That's where pulse width modulation comes in. Let's say you want the truck to go full speed or half speed. Our circuit can only turn the motor on or off, though. With pulse width modulation, we turn the motor on, then turn it off, then turn it on, and so forth. By doing this quickly, we get the truck to go at about half of its top speed. To get it to go a quarter of the top speed, we turn the motor on for a short time—say 0.1 seconds —then turn it off for 0.3 seconds.

Drag and other mechanical losses will force us to adjust these values a bit. Also, the limitations of the Bluetooth low energy connection and constantly redrawing a cool user interface will prevent exceptionally fine control. Still, our software will be able to control the truck quite well.

State engines

There are several ways to write the code that will power and steer the truck using pulse width modulation. Most are very tedious and very hard to write and change. We're going to opt for one that is a little tedious to set up, but very, very easy to tune and modify. It's called a *state engine*, and it's a great trick for your algorithms tool bag. It takes a little thought to wrap your noodle around the idea the first time, but once you get the concept, state engines are both powerful and simple to implement and debug.

Our problem is how to provide power to a motor in several incremental steps. We'll turn the motor on or off over a series of short time intervals—say, 0.1 seconds. For convenience, let's divide the time slices into six 0.1-second increments. We'll designate forward as 1, idle as 0, and reverse as –1. With these definitions, one state is forward motion at full speed. The settings for moving forward at full speed for our six 0.1-second time slices are:

```
1, 1, 1, 1, 1, 1
```

Moving forward at half speed looks like this:

```
1, 0, 1, 0, 1, 0
```

Mapping out several states and encoding these as a techBASIC array constant, we get:

```
DIM stateMap(-4 TO 4, 6)
stateMap = [[-1, -1, -1, -1, -1, -1],
            [-1, -1,  0, -1, -1,  0],
            [-1,  0, -1,  0, -1,  0],
            [-1,  0,  0, -1,  0,  0],
            [ 0,  0,  0,  0,  0,  0],
            [ 1,  0,  0,  1,  0,  0],
            [ 1,  0,  1,  0,  1,  0],
            [ 1,  1,  0,  1,  1,  0],
            [ 1,  1,  1,  1,  1,  1]]
```

Let's take a moment to digest what this table represents and how it is coded in BASIC.

First, the DIM statement looks a little odd. By default, BASIC arrays index from 1 to the largest index specified. The first index of the array is dimensioned as -4 TO 4, though. This sets up an array with nine rows whose indices are –4 to 4. We'll use these as the speed states for the motor, with –4 being full reverse and 4 being full forward. Naturally enough, 0 is idle. The second index of the array tells us that each row of the array will have six time slices, numbered 1 to 6.

The remaining lines initialize the array using a handy, easy to read shortcut. Reading the top row, we see that the motor will be in full reverse for all time slices when the state is –4. When the state is –3, the motor will spend one-third of its time idling and two-thirds of its time applying reverse power. The rest of the states gradually lower the power, and then gradually increase it in the forward direction.

At this point, it's easy to see what the software will do, at least in concept. There will be some way of deciding whether we want the truck to go forward or backward, and how fast. This will be encoded as a number from –4 to 4. The program will loop from 1 to 6 continuously in 0.1-second time slices. It will look at the stateMap array for each time slice, retrieving a value based on the speed state and the index of the current time slice. If the value is –1, the motor will be set to run in reverse; if it is 0, the motor will be told to idle; if the state is 1, the motor will be told to turn forward. This process repeats indefinitely. All the GUI software needs to do is translate the user inputs into a new speed state for each motor and store the value somewhere so the time slice loop knows what state to use.

Back to the Software

The software itself uses the internal accelerometer you learned about back in Chapter 1 to control the speed and direction of the truck. As you tip your iPhone or iPad away from you, the arrow will change from white to successively darker shades of blue, giving visual feedback about the speed. Tipping the iPhone toward you powers the truck in reverse, again with a visual color indicator on the reverse arrow. Tipping the iPhone left or right turns the truck. It's a very simple but intuitive GUI (see Figure 8-27).

Let's take a look at the code.

```
! This app uses the accelerometer to control a car hacked to use the
! RedBear BLE Shield and an Arduino.

redBearUUID$ = "713D0000-503E-4C75-BA94-3148F18D941E"
txUUID$ = "713D0003-503E-4C75-BA94-3148F18D941E"

DIM BLEShield AS BLEPeripheral
DIM txCharacteristic AS BLECharacteristic
```

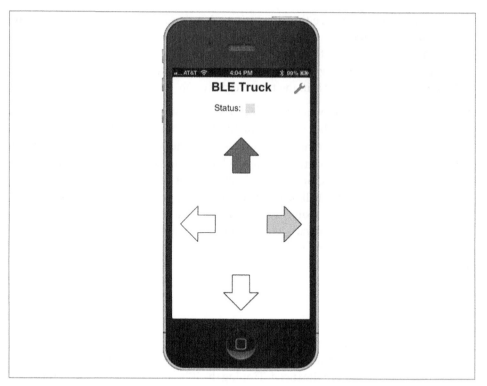

Figure 8-27. The BLE Truck app

The program starts by setting up the UUIDs for the Bluetooth low energy interface to the RedBearLab BLE Shield. The redBearUUID is the UUID for the serial I/O service, and is used to scan for the shield. For this application, all we need to do is transmit information to the device, so the only other UUID we need to define is for the serial transmission characteristic, txUUID. We also set up a couple of variables to hold these so the objects don't get cleaned up between subroutine calls.

You may want to do other things with the RedBearLab shield later. The UUIDs are a bit hard to locate: you can find the complete list on the RedBearLab forum (*https://redbearlab.zendesk.com/entries/22953036-Bluetooth-Protocol-description-*).

```
DIM status AS Label
DIM lastTime AS DOUBLE
```

The next two variables, status and lastTime, hold GUI elements for the status indicator. Just like with the rocket data collection program from Chapter 7, we'll implement

a red-yellow-green label to indicate if the Bluetooth low energy connection is disconnected, attempting to connect, or connected.

```
haveConnection = 0  ❶
turn = 0  ❷
speed = 0  ❸
orientation = 0  ❹
oldPout = 0  ❺
```

The next few variables are used to track various values as the program runs:

❶ haveConnection is set to 1 if there is a valid connection to the Bluetooth low
energy device, and 0 otherwise. It's used to prevent updating the GUI or trying
to send Bluetooth low energy commands if there is no connection.

❷ turn tracks the turn state. Its value will range from -4 to 4, corresponding to the
nine states in the state engine.

❸ Similarly, speed tracks the state of the speed, again ranging from -4 to 4.

❹ orientation records how the user was holding the iPhone or iPad when the
program started. The program will lock in this orientation. This variable is used
when updating the arrows; it's used in the drawArrow subroutine to decide where
the arrows should be drawn and how big they should be.

❺ It takes time to redraw arrows, and sending commands to the RedBearLab shield
uses battery power on both the iPhone and the shield. oldPout is used to keep
track of the last command sent to the shield. Nothing gets updated unless the
command changes.

```
! Set up the speed and turn state engines.
state = 1
maxState = 6
DIM speedForState(-4 TO 4, maxState), turnForState(-4 TO 4, maxState)
speedForState = [[ 4,   4,   4,   4,   4,   4],
                 [ 4,   4,   0,   4,   4,   0],
                 [ 4,   0,   4,   0,   4,   0],
                 [ 4,   0,   0,   4,   0,   0],
                 [ 0,   0,   0,   0,   0,   0],
                 [ 8,   0,   0,   8,   0,   0],
                 [ 8,   0,   8,   0,   8,   0],
                 [ 8,   8,   0,   8,   8,   0],
                 [ 8,   8,   8,   8,   8,   8]]
 turnForState = [[32,  32,  32,  32,  32,  32],
                 [32,  32,   0,  32,  32,   0],
                 [32,   0,  32,   0,  32,   0],
                 [32,   0,   0,  32,   0,   0],
                 [ 0,   0,   0,   0,   0,   0],
                 [16,   0,   0,  16,   0,   0],
                 [16,   0,  16,   0,  16,   0],
```

```
            [16, 16,  0, 16, 16,  0],
            [16, 16, 16, 16, 16, 16]]
```

This chunk of code sets up the state engines. The initial time slice is set to 1, and the maximum number of time slice states is set to 6. This makes it a little easier to change the number of time slices in case you want to experiment a bit. The rest of the program uses the maxState variable to decide how many time states exist, so changing the variable and the two matrices below is all that is needed to update the program to use a different number of time slices.

The two matrices correspond to the state machine matrix discussed previously. The only real difference is using the values 4, 8, 32, and 16 rather than –1 and 1. These are the bitmap values for the byte that will be sent to the Arduino to control the H bridge. All the program has to do is perform a logical or of the appropriate value from each table to create the proper command. All you have to do to change the way the truck performs is to change the numbers in the table.

```
! Draw the GUI.
setUp
```

The last step is to call the setUp subroutine to create the user interface and initiate communications with the Arduino. Let's take a look at this subroutine next, even though it's not sequentially the next one in the program:

```
! Do program setup by drawing the GUI and beginning the scan for the
! RedBear BLE Shield.

SUB setUp
! Switch to the graphics screen.
System.showGraphics(1)
System.setAllowedOrientations(1 << (System.orientation - 1))
orientation = System.orientation
```

After switching on the graphics screen, the program looks to see how the user is holding the iPhone, then locks in that orientation so it won't change. After all, the user is going to be twisting the iPhone all about to control the truck, and the screen should not rotate as the iPhone is tipped.

```
! Get the size of the display.
height = Graphics.height
width = Graphics.width

! Label the app.
DIM title AS Label
title = Graphics.newLabel(10, 10, Graphics.width - 20, 25)
title.setText("BLE Truck")
title.setAlignment(2)
title.setFont("Arial", 28, 1)
```

This puts the name of the program centered at the top of the screen.

```
! Add a status indicator. The status label is set to red initially,
! turns yellow when a connection is made, and changes to green
! when the transmit service is available.
DIM statusLabel AS Label
y = 60
statusLabel = Graphics.newLabel(0, y, width/2)
statusLabel.setText("Status:")
statusLabel.setAlignment(3)
statusLabel.setFont("Arial", 20, 0)

status = Graphics.newLabel(width/2 + 10, y, 21)
status.setBackgroundColor(1, 0, 0)
```

This section of code sets up the red-yellow-green status indicator. It's literally a copy/paste from the Rocket Data program from Chapter 7. It worked great there—why reinvent it here?

```
! Add four direction arrows that will update to give visual feedback
! as the device is tilted.
drawArrow(1, 0)
drawArrow(2, 0)
drawArrow(3, 0)
drawArrow(4, 0)
```

The last step in setting up the user interface is to draw the four arrows. We'll take a look at the drawArrow subroutine shortly.

```
! Start the accelerometer, sampling 10 times a second.
Sensors.setAccelRate(0.1)
```

The program uses the accelerometer for user input. With the UI set up, it's time to start it.

```
! Find the BLE Shield.
BLE.startBLE
DIM uuid(1) AS STRING
uuid(1) = redBearUUID$
BLE.startScan(uuid)
END SUB
```

It would be a Very Bad Thing if the BLE shield started getting signals from the program before the user could see the user interface. We avoid that problem by placing the code to start the Bluetooth low energy connection after the code that starts the user interface.

The drawArrow subroutine is used to draw the four direction arrows that change color as the iPhone is tipped. The first parameter indicates which of the arrows should be updated, while the value parameter indicates the shade of blue. The shade is a number from 0 (white) to 1 (pure blue).

```
! Draw an indicator arrow.
!
! Parameters:
!     direction - The direction of the arrow, which is also used to determine
```

```
!   its position. Directions are one of:
!                 1 - Up
!                 2 - Left
!                 3 - Right
!                 4 - Down
!   value - The value of the arrow, from 0 to 1. This indicates the
!       relative force applied in the direction, where 0 means none and
!       1 means the motor is full power in the given direction.

SUB drawArrow (direction, value)
! Find the size of an arrow.
width = Graphics.width
height = Graphics.height
IF width < height THEN
  arrowSize = width/4
ELSE
  arrowSize = height/4
END IF
top = height/4
border = arrowSize/4
```

The first step is to decide how big the arrows should be based on the screen size.

```
! Find the polygon outlining the arrow.
DIM poly(7, 2)
IF direction = 1 OR direction = 4 THEN
  IF direction = 1 THEN
    y0 = top
    y1 = top + arrowSize/2
    y2 = top + arrowSize
  ELSE
    y0 = height - border
    y1 = y0 - arrowSize/2
    y2 = y0 - arrowSize
  END IF
  x0 = (width - arrowSize)/2
  x1 = x0 + arrowSize/4
  x2 = x0 + arrowSize/2
  x3 = x0 + 3*arrowSize/4
  x4 = x0 + arrowSize
  poly(1, 1) = x2 : poly(1, 2) = y0
  poly(2, 1) = x0 : poly(2, 2) = y1
  poly(3, 1) = x1 : poly(3, 2) = y1
  poly(4, 1) = x1 : poly(4, 2) = y2
  poly(5, 1) = x3 : poly(5, 2) = y2
  poly(6, 1) = x3 : poly(6, 2) = y1
  poly(7, 1) = x4 : poly(7, 2) = y1
ELSE
  IF direction = 2 THEN
    x0 = border
    x1 = x0 + arrowSize/2
    x2 = x0 + arrowSize
  ELSE
```

```
      x0 = width - border
      x1 = x0 - arrowSize/2
      x2 = x0 - arrowSize
   END IF
   y0 = top + (height - top -  border - arrowSize)/2
   y1 = y0 + arrowSize/4
   y2 = y0 + arrowSize/2
   y3 = y0 + 3*arrowSize/4
   y4 = y0 + arrowSize
   poly(1, 1) = x0 : poly(1, 2) = y2
   poly(2, 1) = x1 : poly(2, 2) = y4
   poly(3, 1) = x1 : poly(3, 2) = y3
   poly(4, 1) = x2 : poly(4, 2) = y3
   poly(5, 1) = x2 : poly(5, 2) = y1
   poly(6, 1) = x1 : poly(6, 2) = y1
   poly(7, 1) = x1 : poly(7, 2) = y0
END IF
```

The actual arrows are polygons. In computer graphics, a polygon is made up of a series of lines, where the last line always ends where the first began. It's a bit tedious to set up the polygons themselves, which is all this chunk of code does, but it's not terribly complicated.

```
! Fill the arrow.
Graphics.setColor(1 - value, 1 - value, 1)
Graphics.fillPoly(poly)

! Outline the arrow in black.
Graphics.setColor(0, 0, 0)
Graphics.drawPoly(poly)
END SUB
```

There are two basic operations with a polygon: filling it and outlining it. We do both—but it's important to fill it with the appropriate shade of blue before drawing the outline, or the fill color will obliterate the outline.

Setting up the Bluetooth low energy connection follows the same familiar pattern seen in the previous two chapters:

```
! Called when a peripheral is found. If it is a RedBear BLE shield, we
! initiate a connection to it and stop scanning for peripherals.
!
! Parameters:
!    time - The time when the peripheral was discovered.
!    peripheral - The peripheral that was discovered.
!    services - List of services offered by the device.
!    advertisements - Advertisements (information provided by the
!        device without the need to read a service/characteristic)
!    rssi - Received Signal Strength Indicator

SUB BLEDiscoveredPeripheral (time AS DOUBLE,
                              peripheral AS BLEPeripheral,
```

```
                              services() AS STRING,
                              advertisements(,) AS STRING,
                              rssi)
    BLE.connect(peripheral)
    BLE.stopScan
    BLEShield = peripheral
    haveConnection = 1
    END SUB
```

Once the RedBearLab BLE Shield is found, the program records the peripheral and
stops scanning.

```
    ! Called to report information about the connection status of the
    ! peripheral or to report that services have been discovered.
    !
    ! Parameters:
    !    time - The time when the information was received.
    !    peripheral - The peripheral.
    !    kind - The kind of call. One of
    !          1 - Connection completed
    !          2 - Connection failed
    !          3 - Connection lost
    !          4 - Services discovered
    !    message - For errors, a human-readable error message.
    !    err - If there was an error, the Apple error number. If there
    !          was no error, this value is 0.

    SUB BLEPeripheralInfo (time AS DOUBLE,
                           peripheral AS BLEPeripheral,
                           kind AS INTEGER,
                           message AS STRING,
                           err AS LONG)
    DIM uuid(1) AS STRING
    IF kind = 1 THEN ❶
      ! The connection was established. Discover the service.
      uuid(1) = redBearUUID$
      peripheral.discoverServices(uuid)
      status.setBackgroundColor(1, 1, 0): ! Connection made: Status Yellow.
    ELSE IF kind = 2 OR kind = 3 THEN ❷
      ! Lost the connection--Change the status and begin looking again.
      status.setBackgroundColor(1, 0, 0): ! Connection lost: Status Red.
      haveConnection = 0
      BLE.connect(peripheral)
    ELSE IF kind = 4 THEN ❸
      ! Once the RedBear service is found, start discovery on the characteristics.
      DIM availableServices(1) AS BLEService
      availableServices = peripheral.services
      FOR a = 1 TO UBOUND(availableServices, 1)
        IF availableServices(a).UUID = redBearUUID$ THEN
          uuid(1) = txUUID$
          peripheral.discoverCharacteristics(uuid, availableServices(a))
        END IF
      NEXT
```

```
END IF
END SUB
```

This subroutine responds in three different ways, depending on why it was called:

❶   Once a connection is established, start the process of discovering services and change the status indicator to red.

❷   If the connection is lost, change the status indicator to red, set the haveConnec tion flag to false, and start scanning again.

❸   Once services are discovered, record the serial transmit service and begin discovery of characteristics.

```
! Called to report information about a characteristic or included
! services for a service. If it is one we are interested in, start
! handling it.
!
! Parameters:
!    time - The time when the information was received.
!    peripheral - The peripheral.
!    service - The service whose characteristic or included
!        service was found.
!    kind - The kind of call. One of
!        1 - Characteristics found
!        2 - Included services found
!    message - For errors, a human-readable error message.
!    err - If there was an error, the Apple error number. If there
!        was no error, this value is 0.

SUB BLEServiceInfo (time AS DOUBLE,
                    peripheral AS BLEPeripheral,
                    service AS BLEService,
                    kind AS INTEGER,
                    message AS STRING,
                    err AS LONG)
IF kind = 1 THEN
  ! Get the characteristics.
  DIM characteristics(1) AS BLECharacteristic
  characteristics = service.characteristics
  FOR i = 1 TO UBOUND(characteristics, 1)
    IF characteristics(i).uuid = txUUID$ THEN
      ! Remember the transmit service.
      txCharacteristic = characteristics(i)

      ! Connection complete: Status Green.
      status.setBackgroundColor(0, 1, 0)
      haveConnection = 1

      ! Set the four pins we use so the Firmata software treats them
      ! as output pins.
      DIM value(3) AS INTEGER
```

```
      value = [$F4, $02, $01]
      FOR pin = 2 TO 5
        BLEShield.writeCharacteristic(txCharacteristic, value)
        value(2) = value(2)*2
      NEXT
    END IF
  NEXT
END IF
END SUB
```

Once the serial transmit service is available, the program is ready to go. It changes the status indicator to green and sets the `haveConnection` flag to true so the GUI will begin updating. There is one new twist in this subroutine. The Firmata software we loaded onto the Arduino can do a number of different things with the various digital I/O pins. This program uses pins 2 through 5 to set the motor state, so these four pins must be set to an output state. The Firmata command to set the pin mode is a three-byte sequence that starts with hexadecimal `$F4`, followed by the pin number and finally by a 1, indicating the pin is an output pin. The `FOR` loop sets pins 2 through 5 to output pins.

With the user interface set up and a connection established to the RedBearLab BLE Shield, the only remaining task is to implement the state engine that will drive the truck. If you've been following along through the other chapters, it will come as no surprise that this is done in the `nullEvent` subroutine, which is called constantly when the program is not doing other tasks:

```
! Check the accelerometer and send appropriate commands to the car.
!
! Parameters:
!    time - The time when the call was made.

SUB nullEvent (time AS DOUBLE)
IF time > lastTime + 0.1 AND haveConnection AND txCharacteristic <> NULL THEN
```

The first step is to make sure the program is ready to handle changes in the state engine.

The first check looks to see if at least 0.1 seconds have elapsed since the last time the state was updated. If not, there is no point in going through the remaining updates.

You've already seen how `haveConnection` is updated. Checking it here makes sure there is a connection. Checking `txCharacteristic` also makes sure the serial output characteristic is available.

```
! Get the angle of the device for speed and turning.
a = Sensors.accel
SELECT CASE orientation
  CASE 1 : ! Home button down
    turnAngle = DEG(ANGLE(a(1), a(3)) + PI/2)
    speedAngle = DEG(ANGLE(a(2), a(3)) + PI/2)

  CASE 2 : ! Home button left
```

```
    speedAngle = -DEG(ANGLE(a(1), a(3)) + PI/2)
    turnAngle = DEG(ANGLE(a(2), a(3)) + PI/2)

  CASE 3 : ! Home button right
    speedAngle = DEG(ANGLE(a(1), a(3)) + PI/2)
    turnAngle = -DEG(ANGLE(a(2), a(3)) + PI/2)

  CASE 4: ! Home button up
    turnAngle = -DEG(ANGLE(a(1), a(3)) + PI/2)
    speedAngle = -DEG(ANGLE(a(2), a(3)) + PI/2)
END SELECT
```

The first step in updating the system is to check and see how the iPhone is being held. The accelerometer gives all the information needed, returning the acceleration along the x-, y-, and z-axes. ANGLE returns the angle based on two of the values, and DEG converts it from radians to degrees. We need to find the angle four different ways, depending on which of the four orientations the user was holding the iPhone in when the GUI was set up, but each calculation is pretty simple.

```
! Decide on the proper speed.
newSpeed = INT(speedAngle/8)
```

With the speed angle—the forward and backward rotation angle—in hand, we convert from an angle to a motor state. We start by dividing the angle into integral states, where each 8-degree chunk is turned into an integer. Anything from 0 to 8 degrees is changed to 0, 8 to 16 degrees is changed to 1, and so forth. Negative angles are important, too. This step converts values from –8 to 0 to –1, –16 to –8 to –2, and so on.

```
IF newSpeed < 0 THEN newSpeed = newSpeed + 1
```

Adding 1 to speeds from negative angles gives a nice flat range of 16 degrees when the iPhone is roughly level, where the truck is idled.

```
IF newSpeed < -4 THEN newSpeed = -4
IF newSpeed > 4 THEN newSpeed = 4
```

Just in case the user gets really excited and tips the iPhone way over, we'll pin the values to the range –4 to 4. This prevents an array subscript out of range error.

```
IF newSpeed <> speed THEN
  IF newSpeed = 0 THEN
    IF speed > 0 THEN drawArrow(1, 0)
    IF speed < 0 THEN drawArrow(4, 0)
  ELSE IF newSpeed < 0 THEN
    IF speed > 0 THEN drawArrow(1, 0)
    drawArrow(4, -newSpeed/4)
  ELSE
    drawArrow(1, newSpeed/4)
    IF speed < 0 THEN drawArrow(4, 0)
  END IF
  speed = newSpeed
END IF
```

If the speed has changed since the last pass through the loop, it's time to redraw the arrows for forward and reverse. The code is careful to only redraw an arrow if it needs to. Redrawing the arrow doesn't take much time, but updating a large portion of the graphics screen can—especially on the iPad, where the arrows are fairly large.

```
! Decide on the proper direction.
newTurn = INT(turnAngle/8)
IF newTurn < 0 THEN newTurn = newTurn + 1
IF newTurn < -4 THEN newTurn = -4
IF newTurn > 4 THEN newTurn = 4
IF newTurn <> turn THEN
  IF newTurn = 0 THEN
    IF turn < 0 THEN drawArrow(2, 0)
    IF turn > 0 THEN drawArrow(3, 0)
  ELSE IF newTurn < 0 THEN
    drawArrow(2, -newTurn/4)
    IF turn > 0 THEN drawArrow(3, 0)
  ELSE
    IF turn < 0 THEN drawArrow(2, 0)
    drawArrow(3, newTurn/4)
  END IF
  turn = newTurn
END IF
```

Here the program does it all again for turning. The code exactly parallels the code for forward and reverse.

```
! Advance the state.
state = state + 1
IF state > maxState THEN state = 1
```

The state engine's time state gets updated.

```
! Decide whether each motor should be forward, off or reversed.
pout = speedForState(speed, state)
pout = pout BITOR turnForState(turn, state)
IF oldPout <> pout THEN
  oldPout = pout

  ! Send the command to the BLE shield.
  DIM value(3) AS INTEGER
  value = [$90, pout, $00]
  BLEShield.writeCharacteristic(txCharacteristic, value, 0)
END IF
```

Assuming there has been a change, the new motor speeds are written to the RedBearLab shield. Like the Firmata command used to set the pin state that was in the BLEServi ceInfo subroutine, this command is a three-byte command. The first byte of $90 tells Firmata to set the state of digital output pins 0 through 14. On the Arduino Uno, these are the 15 pins numbered 0 to 14 along the right edge of the board. The next two bytes are a bitmap that tells Firmata whether each pin should be set to 0 or 1. For example, to set pin 0 to 1 and all other pins to 0, pass 1 in the second byte of the command and 0 in the third byte.

Now you can see where the values used in the state table actually come from. If the value is 4, pin 2 will be turned on. If the value is 8, pin 3 will be turned on. Looking at the state table, you can see that turning pin 2 on will also turn pin 3 off, and vice versa. The second state table uses the values 16 and 32 the same way, controlling pins 4 and 5. The program combines these values using the BITOR operation, so it's possible, for example, to go forward and right at the same time. That's the last piece of the puzzle—you now know how the truck is controlled. You also know how to set bits 2–13 on the Arduino using Bluetooth low energy. The shield hijacks bits 0 and 1 for its own use, though; don't mess with them.

Exploring Firmata

This program shows how to control pins 2–5 on the Arduino, using them for output. It will take a bit of digging in the Firmata code to figure out the details, but it can do a lot more. Pins can be used for digital input, too, or for analog input or output. You can even control servos.

This program told the software to treat pins 2–5 as digital output using the $F4 command, then used the $90 command to actually write to those pins. Other commands are used for the other functions available through Firmata. To explore, start with an Internet search for "arduino firmata."

```
    ! Remember the time so we don't do this constantly.
    lastTime = time
  END IF
END SUB
```

Finally, the time is updated so the 0.1-second interval before the next pass through the loop is restarted.

Start Your Engines!

That was a lot of work. It's time for some fun. Flip the switch on, making sure you get power to the Arduino. An LED will illuminate if there is power. Run BLE Truck from a Bluetooth low energy–capable iPhone, iPad, or iPod Touch. Tilt it so it is flat as the connection is established. Once the connection indicator is green, give it a tilt forward. Make sure the truck really goes forward. If not, flip the Molex connector running to the front motor. Repeat the test with the steering motor.

Once you are sure the connectors are connected in the proper direction, reattach the truck body using the four screws that you carefully saved when disassembling the truck oh so many pages ago.

Now go find a cat.

CHAPTER 9

Peer-to-Peer Bluetooth Low Energy

About This Chapter

Prerequisites

Read Chapter 6 first to get a basic understanding of Bluetooth low energy technology.

Equipment

You will need two iOS devices. Each must be either an iPhone 4S or later, an iPad 3 or later, or an iPod 5th Gen or later running iOS 6 or later.

Software

You will need a copy of techBASIC or techBASIC Sampler.

What You Will Learn

An iPhone was used to connect to an external Bluetooth low energy device in the three previous chapters. This chapter reverses the connection. You will see how to make the iPhone itself become the Bluetooth low energy device, broadcasting information for Bluetooth low energy–compatible controllers to see. We'll put this to use to create a program that performs peer-to-peer communications between two iOS devices.

Bluetooth low energy is normally limited to short, 20-byte packets of information. The program in this chapter also shows how to send longer files.

Bluetooth Low Energy Slave Mode

So far, all of our Bluetooth low energy projects have used the iPhone or iPad to access peripherals that offer information to any Bluetooth low energy device. The peripheral, like the Texas Instruments SensorTag or the RedBearLab BLE Shield, is called the *slave* device, while the consumer of the information—in our case, the iPhone or iPad—is called the *central* device.

But what if we want things to go the other way? What if we want the iPhone to be the peripheral, sending information to other devices that subscribe to its service? That's what this chapter is all about.

Since this is an iPhone and iPad book, we'll explore Bluetooth low energy slave mode with a program that lets two iPhones or iPads talk to each other. You will need two Bluetooth low energy–capable devices. If you only have one, you can still scan the chapter to see how to create slave devices, then access the information served by the iPhone using another Bluetooth low energy device (perhaps one of the newer Macintosh computers).

BLE Chat

BLE Chat is a peer-to-peer terminal program (see Figure 9-1). Once two devices are connected, anything typed on one device is sent to the other, and vice versa. Both devices operate as both central and slave devices. It's old-school texting implemented with a modern Bluetooth low energy feature set.

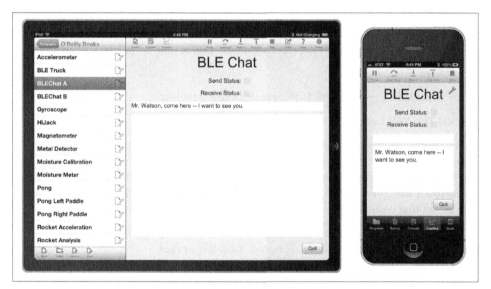

Figure 9-1. User interface for the BLE Chat program

Setting Up the Services

From working through the Bluetooth low energy projects in the last few chapters, you should be familiar with the way Bluetooth low energy communication works. A

Bluetooth low energy slave broadcasts an advertisement that says, "Here I am, and here's what I can do." Our programs then look for one or more *services* that we need—an accelerometer, perhaps. The services contain one or more *characteristics*, each of which acts like a variable that can be read, written, or both. We'll need to create all of these to set up our own Bluetooth low energy slave.

Here's the subroutine from BLE Chat that sets up the service and sends the advertisement. You'll find the complete program in the *O'Reilly Books* folder in techBASIC and techBASIC Sampler:

```
! Set up the communication service and advertise.

SUB advertise
peripheralManager = BLE.newBLEPeripheralManager ❶

DIM service AS BLEMutableService ❷
service = peripheralManager.newService(localServiceUUID$, 1)

localTextCharacteristic = service.newCharacteristic(localTextCharacteristicUUID$,
                                                    $0012, $0001) ❸
localReadyCharacteristic = service.newCharacteristic(localReadyCharacteristicUUID$,
                                                     $0006, $0003)

peripheralManager.addService(service) ❹
peripheralManager.startAdvertising(localName$) ❺
END SUB
```

Here's what's going on in this subroutine:

❶ The services for a Bluetooth low energy slave device are managed by a class called BLEPeripheralManager. We'll need one of them to set up our services. This line gets a BLEPeripheralManager from the predefined BLE class.

❷ Next, the program creates a new service. Instead of the BLEService you've seen in previous programs, BLEMutableService is used to create the slave service. This is a general pattern throughout the various classes used to set up a Bluetooth low energy slave. Each of the classes used to access a Bluetooth low energy device has a corresponding class with "mutable" added to the name. That's a convention Apple adopted, and techBASIC sticks with it. Mutable just means the object can change, while immutable means it cannot. We're setting up Bluetooth low energy services that can change, so it makes sense to put Mutable in the names of these classes to distinguish them from the central classes that are immutable.

The first parameter gives the UUID for the service, while the second indicates that this is a primary service for the device.

❸ Our service has two characteristics. The first, `localTextCharacteristic`, is used to send text to any other device. The characteristic has its own UUID. The other parameters will be explained in a moment. The second service, `localRea dyCharacteristic`, will be used for handshaking, a topic we'll explore in detail as we work through the code.

❹ Now that the characteristics are defined, the service is added to the peripheral manager.

❺ The device begins advertising that it has a service to offer. Other devices can now find and connect to this one.

`BLEMutableService.newCharacteristic` has three parameters. The first is pretty obvious—any characteristic needs a UUID to uniquely identify the characteristic. The next two describe the precise kind of characteristic that this is.

The second parameter is a bitmapped value, where each bit sets one capability for the service. The parameter of $0012 sets two bits, one indicating the characteristic supports notifications and the other saying it can be read. If you recall from the SensorTag application, characteristics that support notifications push new values out when they are available—exactly what we need in this case. The value of $0006 says that the characteristic can be read or written, it does not support notification. Other possible values are described in the techBASIC Reference Manual (*http://www.byteworks.us/ Byte_Works/Documentation.html*) and online help system.

The last parameter says the value can be read without using encryption ($0001) or read and written without encryption ($0003). It's also possible to set up secure transmissions between devices, but we'll keep things simple by assuming our chats won't need protection from the NSA.

Using the Services

It was deceptively simple to set up the Bluetooth low energy stack. Let's look at the rest of the program to see how it is used.

```
! BLE Chat sets up peer-to-peer communication between two BLE-equipped
! iOS devices. Run BLE Chat A on one device, and BLE Chat B on the other.

! Set up global GUI controls.
DIM quitButton AS Button, console AS TextView, inputLine AS TextField
DIM sendStatus AS Label, receiveStatus AS Label
```

The program starts with the familiar declarations of some GUI controls. These are the ones that require access from more than one subroutine.

```
! Set up a variable to hold any text that has not yet been sent to the
! central device.
line$ = ""
```

Bluetooth low energy is designed for short data packets of no more than 20 bytes. Our program wants to send messages that will often exceed that limit, so we'll have to design a mechanism that gets around this limitation. We'll see how this is done later. The global variable line$ is used by the code that manages sending the data in chunks.

```
! Create variables to hold the peripherals and characteristics.
DIM localTextCharacteristic AS BLEMutableCharacteristic
DIM localReadyCharacteristic AS BLEMutableCharacteristic
DIM remotePeripheral AS BLEPeripheral
DIM remoteReadyCharacteristic AS BLECharacteristic
DIM peripheralManager AS BLEPeripheralManager
```

There are two versions of the chat program. This isn't strictly necessary, but it shows a framework that allows setting up a single central device that can communicate with multiple peripheral devices, acting like a server that can relay information from one device to another. To help keep all of this straight, the program uses a simple naming convention. Variables that start with local hold information about the service created on the device the program is running on. Variables that start with remote hold information about the other device—the one we're chatting with.

```
! Set up the UUIDs and names.
isA = 1
```

This dual role means there are two versions of the chat program, called BLEChatA and BLEChatB. The only difference between the two programs is the setting for the isA variable. One is set to 1, and the other to 0.

```
IF isA THEN
  localName$ = "BLEChatA"
  localServiceUUID$ = "01A00C7B-153C-45F9-B083-FE135E4E5CA0"
  localTextCharacteristicUUID$ = "01A10C7B-153C-45F9-B083-FE135E4E5CA0"
  localReadyCharacteristicUUID$ = "01A20C7B-153C-45F9-B083-FE135E4E5CA0"
  remoteName$ = "BLEChatB"
  remoteServiceUUID$ = "01B00C7B-153C-45F9-B083-FE135E4E5CA0"
  remoteTextCharacteristicUUID$ = "01B10C7B-153C-45F9-B083-FE135E4E5CA0"
  remoteReadyCharacteristicUUID$ = "01B20C7B-153C-45F9-B083-FE135E4E5CA0"
ELSE
  localName$ = "BLEChatB"
  localServiceUUID$ = "01B00C7B-153C-45F9-B083-FE135E4E5CA0"
  localTextCharacteristicUUID$ = "01B10C7B-153C-45F9-B083-FE135E4E5CA0"
  localReadyCharacteristicUUID$ = "01B20C7B-153C-45F9-B083-FE135E4E5CA0"
  remoteName$ = "BLEChatA"
  remoteServiceUUID$ = "01A00C7B-153C-45F9-B083-FE135E4E5CA0"
  remoteTextCharacteristicUUID$ = "01A10C7B-153C-45F9-B083-FE135E4E5CA0"
  remoteReadyCharacteristicUUID$ = "01A20C7B-153C-45F9-B083-FE135E4E5CA0"
END IF
```

The program then sets up its own name and the UUIDs for the service and characteristic generated in the advertise subroutine. It expects a different name and UUID from the program it will connect to. Of course, they are mirror images of one another.

It's fair to ask where these UUIDs came from. The answer is that I made them up with the help of a random number generator. Remember, we're building a set of Bluetooth low energy services, so it's up to us to define the UUIDs the device will use to tell the world it has a service to provide.

Doesn't that mean it's possible to unintentionally create two devices with the same UUIDs, causing problems if they ever get close to one another? Well, yes, but it's pretty unlikely. That's why UUIDs that are not set aside for industry standard uses, like a heart rate monitor, are 128 bits long. This makes the chances of an accidental collision of UUIDs about 1 in 3.4×10^{38}. If you tried a billion UUIDs a second, it would take far, far longer than the lifetime of the universe to get an accidental collision of UUIDs. In statistical mechanics, that's a working definition for impossible. In programming, it means I'll get two or three bug reports that I never figure out.

To keep things simple, the family of UUIDs needed for this program was created by varying two hexadecimal digits from the original, random UUID.

```
! Establish communication and set up the UI.
advertise
scanForChats
setUpGUI
```

With the global variables defined, the program calls advertise to set up its own service, starts scanning for the other device, and sets up the user interface. We've already seen the first of these subroutines. Here's the one that looks for the other device:

```
! Start scanning for another device to talk to.

SUB scanForChats
BLE.startBLE
DIM uuid(0) AS STRING
BLE.startScan(uuid)
END SUB
```

This is the same code you've seen in previous Bluetooth low energy programs. It's just captured in a subroutine to make it easier to see which code is used to provide a service and which code is used to look for one on another device.

As with previous programs, there are a series of subroutines that are used to actually get information from the other device.

```
! Called when a peripheral is found. If it is another chat client, we
! initiate a connection to it and stop scanning for peripherals.
!
! Parameters:
!    time - The time when the peripheral was discovered.
!    peripheral - The peripheral that was discovered.
!    services - List of services offered by the device.
!    advertisements - Advertisements (information provided by the
!        device without the need to read a service/characteristic)
```

```
!    rssi - Received Signal Strength Indicator

SUB BLEDiscoveredPeripheral (time AS DOUBLE, _
                             peripheral AS BLEPeripheral, _
                             services() AS STRING, _
                             advertisements(,) AS STRING, _
                             rssi)
  FOR i = 1 TO UBOUND(advertisements, 1)
    IF advertisements(i, 1) = "kCBAdvDataLocalName" THEN
      IF advertisements(i, 2) = remoteName$ THEN
        BLE.connect(peripheral)
        remotePeripheral = peripheral
      END IF
    END IF
  NEXT
END SUB
```

BLEDiscoveredPeripheral is called when a device is found. It checks to make sure it's the device we want and, if so, tries to connect with the device.

```
! Called to report information about the connection status of the
! peripheral or to report that services have been discovered.
!
! Parameters:
!    time - The time when the information was received.
!    peripheral - The peripheral.
!    kind - The kind of call. One of
!        1 - Connection completed
!        2 - Connection failed
!        3 - Connection lost
!        4 - Services discovered
!    message - For errors, a human-readable error message.
!    err - If there was an error, the Apple error number. If there
!        was no error, this value is 0.

SUB BLEPeripheralInfo (time AS DOUBLE, _
                       peripheral AS BLEPeripheral, _
                       kind AS INTEGER, _
                       message AS STRING, _
                       err AS LONG)
  DIM uuid(0) AS STRING
  SELECT CASE kind
    CASE 1
      peripheral.discoverServices(uuid)
      BLE.stopScan
      sendStatus.setBackgroundColor(1, 1, 0): ! Connection made: Status Yellow.

    CASE 2, 3
      BLE.startScan(uuid)
      sendStatus.setBackgroundColor(1, 0, 0): ! Connection lost: Status Red.

    CASE 4
      DIM services(1) AS BLEService, included(1) AS BLEService
```

```
        services = peripheral.services
        FOR i = 1 TO UBOUND(services, 1)
          IF services(i).uuid = remoteServiceUUID$ THEN
            peripheral.discoverCharacteristics(uuid, services(i))
          END IF
        NEXT
    END SELECT
    END SUB
```

BLEPeripheralInfo handles the connection with the device. As with our previous programs, if kind is 1, the connection is established, so we start looking for services. A kind of 4 indicates we've found one. The program checks to see if it's the one we want and, if so, starts looking for characteristics. This subroutine also handles setting the status label color as the connection is made and lost.

```
! Called to report information about a characteristic or included
! services for a service. If it is one we are interested in, start
! handling it.
!
! Parameters:
!    time - The time when the information was received.
!    peripheral - The peripheral.
!    service - The service whose characteristic or included
!        service was found.
!    kind - The kind of call. One of
!        1 - Characteristics found
!        2 - Included services found
!    message - For errors, a human-readable error message.
!    err - If there was an error, the Apple error number. If there
!        was no error, this value is 0.

SUB BLEServiceInfo (time AS DOUBLE, _
                    peripheral AS BLEPeripheral, _
                    service AS BLEService, _
                    kind AS INTEGER, _
                    message AS STRING, _
                    err AS LONG)
IF kind = 1 THEN
  DIM characteristics(1) AS BLECharacteristic
  characteristics = service.characteristics
  FOR i = 1 TO UBOUND(characteristics, 1)
    IF characteristics(i).uuid = remoteTextCharacteristicUUID$ THEN
      peripheral.setNotify(characteristics(i), 1)
      ! Connection complete: Status Green.
      sendStatus.setBackgroundColor(0, 1, 0):
    ELSE IF characteristics(i).uuid = remoteReadyCharacteristicUUID$ THEN
      remoteReadyCharacteristic = characteristics(i)
    END IF
  NEXT
END IF
END SUB
```

Once the proper characteristic is found, it is set to provide notifications. The remoteR
eadyCharacteristic is saved for later.

```
! Called to return information from a characteristic.
!
! Parameters:
!    time - The time when the information was received.
!    peripheral - The peripheral.
!    characteristic - The characteristic whose information
!        changed.
!    kind - The kind of call. One of
!        1 - Called after a discoverDescriptors call.
!        2 - Called after a readCharacteristics call.
!        3 - Called to report status after a writeCharacteristics
!            call.
!    message - For errors, a human-readable error message.
!    err - If there was an error, the Apple error number. If there
!        was no error, this value is 0.
!

SUB BLECharacteristicInfo (time AS DOUBLE, _
                          peripheral AS BLEPeripheral, _
                          characteristic AS BLECharacteristic, _
                          kind AS INTEGER, _
                          message AS STRING, _
                          err AS LONG)
IF kind = 2 THEN
  ! Get the data and display it in the console view.
  DIM value(1) AS INTEGER
  value = characteristic.value
  text$ = console.getText
  FOR i% = 1 TO UBOUND(value, 1)
    text$ = text$ & CHR(value(i%))
  NEXT
  console.setText(text$)

  ! Tell the slave device we are ready for more.
  value = [1]
  remotePeripheral.writeCharacteristic(remoteReadyCharacteristic, value, 0)
ELSE IF kind = 3 AND err <> 0 THEN
  PRINT "Error writing "; characteristic.uuid; ": ("; err; ") "; message
END IF
END SUB
```

BLECharacteristicInfo handles the notifications, adding any text to the large text view
that appears in the GUI.

It also tells the slave device we're ready for another packet of data by setting the value
of remoteReadyCharacteristic to 1. This is one half of the handshaking needed to
send more than 20 bytes of data from one device to the other.

A simplistic implementation of communications between the two devices might just send the first 20 bytes of text, then send the next 20 bytes, and so forth until all of the text has been sent. Unfortunately, that won't work. The problem is that the Bluetooth low energy communications channel can get jammed with too much information, and packets will get dropped.

To get around this problem, the slave device is going to send the central program 20 bytes of text, then wait until the central program says it has received the data and finished processing it. The slave device sets the value of the remoteReadyCharacteristic characteristic to 0 just before sending a data packet, and won't send more information until it sees the value of remoteReadyCharacteristic changed back to 1 by the central device that is receiving the data. This subroutine is setting the value of remoteReadyCharacteristic to 1 so the slave device knows it is allowed to send more data if it has any.

We'll see the other half of this arrangement in a moment.

```
! Called to return information from a characteristic of
! a peripheral manager.
!
! Parameters:
!    time - The time when the information was received.
!    peripheral - The peripheral.
!    characteristic - The characteristic whose information
!        changed.
!    kind - The kind of call. One of
!        1 - Called after a central subscribes to a characteristic.
!        2 - Called after a central unsubscribes from a characteristic.
!

SUB BLEMutableCharacteristicInfo (time AS DOUBLE, _
                                 peripheral AS BLEPeripheralManager, _
                                 characteristic AS BLECharacteristic, _
                                 kind AS INTEGER)
IF kind = 1 THEN
  receiveStatus.setBackgroundColor(0, 1, 0): ! Subscription active: Status Green.
ELSE
  receiveStatus.setBackgroundColor(1, 0, 0): ! Subscription inactive: Status Red.
END IF
END SUB
```

All of the Bluetooth low energy subroutines we've just looked at up to this one should be pretty familiar by now, since they have appeared in every Bluetooth low energy program so far. There is one Bluetooth low energy subroutine that is new to slave mode, though: BLEMutableCharacteristicInfo is called by the operating system when a remote Bluetooth low energy central device subscribes to one of our services, asking for notifications. This tells the slave program that it's time to start sending updates to the values—perhaps by turning on an accelerometer. The same call is used if the central device decides to stop listening to the service, letting the program turn off power-hungry sensors.

Our use is a bit more mundane, but still important. This implementation sets the color of the receiveStatus label, giving the user a visual cue to let him know if the other device is ready for any text we might want to send.

```
! Set up the user interface.

SUB setUpGUI
Graphics.setColor(0.9, 0.9, 0.9)
Graphics.fillRect(0, 0, Graphics.width, Graphics.height)

DIM title AS Label
title = Graphics.newLabel(20, 20, Graphics.width - 40, 40)
title.setFont("Sans-serif", 48, 0)
title.setText("BLE Chat")
title.setAlignment(2)
title.setBackgroundColor(0, 0, 0, 0)

DIM sendStatusLabel AS Label
x = Graphics.width/2 - 100
y = 90
sendStatusLabel = Graphics.newLabel(x, y, 150)
sendStatusLabel.setText("Send Status:")
sendStatusLabel.setAlignment(3)
sendStatusLabel.setFont("Arial", 20, 0)
sendStatusLabel.setBackgroundColor(0, 0, 0, 0)

sendStatus = Graphics.newLabel(x + 160, y, 21)
sendStatus.setBackgroundColor(1, 0, 0)

DIM receiveStatusLabel AS Label
y = y + 41
receiveStatusLabel = Graphics.newLabel(x, y, 150)
receiveStatusLabel.setText("Receive Status:")
receiveStatusLabel.setAlignment(3)
receiveStatusLabel.setFont("Arial", 20, 0)
receiveStatusLabel.setBackgroundColor(0, 0, 0, 0)

receiveStatus = Graphics.newLabel(x + 160, y, 21)
receiveStatus.setBackgroundColor(1, 0, 0)

y = y + 41
inputLine = Graphics.newTextField(20, y, Graphics.width - 40)
inputLine.setBackgroundColor(1, 1, 1)
inputLine.setFont("Sans_serif", 20, 0)

y = y + 41
console = Graphics.newTextView(20, y, Graphics.width - 40, Graphics.height - 292)
console.setEditable(0)
console.setFont("Sans_serif", 20, 0)
```

```
quitButton = Graphics.newButton(Graphics.width - 92, Graphics.height - 57)
quitButton.setTitle("Quit")
quitButton.setBackgroundColor(1, 1, 1)
quitButton.setGradientColor(0.7, 0.7, 0.7)

System.showGraphics
END SUB
```

setUpGUI sets up the user interface, creating the labels, Quit button, text field, and text view for the app. The text field is used for the input field, while the text view, which can span multiple lines, is used to display the information received from the other device. This should be pretty familiar code by now, so we won't go through it line by line.

```
! Handle a press of the enter key by sending the text to the connected device.
!
! Parameters:
!    ctrl - The text field that changed.
!    time - The time stamp when the enter key was pressed.

SUB valueChanged (ctrl AS Control, time AS DOUBLE)
IF ctrl = inputLine THEN
  line$ = inputLine.getText & CHR(10)
  inputLine.setText("")
  sendText
END IF
END SUB
```

valueChanged is called when the user taps Enter, signaling that the line of text in the text field is ready to send. The text is saved to the global variable line$, and then the subroutine sendText is called.

```
! This utility routine takes the line stored in the global variable line$
! and sends up to 20 characters from the line to the central device. It
! checks to be sure the data was sent, then removes the characters from
! the line.

SUB sendText
! Make sure the central device is ready for more data. If not wait, but
! time out after 0.5 seconds.
time# = System.ticks
DIM value2(1) AS INTEGER
DO
  value2 = localReadyCharacteristic.value
  done = (UBOUND(value2, 1) = 1) AND (value2(1) = 1)
LOOP WHILE (NOT done) AND (System.ticks - time# < 0.5)
```

sendText implements the other half of the handshaking arrangement needed to safely send more than 20 bytes of data from a Bluetooth low energy slave to a central device. The first step is to make sure the other device is ready for data. That would seem to be easy   enough—the   program   could   just   check   the   value   of   the localReadyCharacteristic characteristic to make sure it is 1, then send the bytes. If the value is not 1, the program could loop, waiting until it is safe to send more data.

And that's a recipe for an infinite loop. Ouch.

Sure, it would work most of the time, but what if the other device never sets localRea dyCharacteristic to 1? Maybe it's disconnected, or maybe there was some radio interference. The point is that our program may never see localReadyCharacteristic set to 1. That's why the code is checking to see if 0.5 seconds have elapsed with no response from the central device. If that happens, we push on and hope for the best.

It's not ideal, and in some applications you might use a different strategy, but it sure beats a program that hangs in an infinite loop.

```
! Place up to 20 bytes into a value array.
IF LEN(line$) > 20 THEN
  length% = 20
ELSE
  length% = LEN(line$)
END IF
DIM value(length%) AS INTEGER
FOR i% = 1 TO length%
  value(i%) = ASC(MID(line$, i%, 1))
NEXT
```

This little chunk of code gets up to 20 characters, converts them to bytes, and stuffs them into an array.

```
! Set our handshaking value to 0. The central will set it to 1 after
! receiving the data.
value2 = [0]
localReadyCharacteristic.setValue(value)
```

Just before sending the chunk of data, the program sets localReadyCharacteristic to 0. As you saw a moment ago, the central device will set it back to 1--we hope—after the data packet is processed.

```
! Send the data to the central device.
result% = peripheralManager.updateValue(localTextCharacteristic, value)
```

Finally, we call updateValue to send the packet of data to the central device.

This call can fail, too. If we send packets too quickly, the local operating system may return right away, telling us that the communication channel is full. If result% is set to 1, everything is fine. If result% is set to 0, the data could not be sent. In that case, the operating system will call a subroutine named readyToUpdateSubscribers as soon as the channel is available. The data can be resent at that time.

```
! If the send was successful, remove the bytes from the line.
IF result% THEN
  console.setText(console.getText & LEFT(line$, 20))
```

```
      IF LEN(line$) > 20 THEN
        line$ = RIGHT(line$, LEN(line$) - 20)
        sendText
      ELSE
        line$ = ""
      END IF
    END IF
  END SUB
```

Since the write can fail, sendText checks to make sure it succeeded before removing the text that was just sent from the line$ variable. If the call to updateValue did fail, all of the text is still sitting in line$, patiently waiting for the next call to sendText.

```
! An attempt was made to send data to the central device, but the I/O channel
! was busy. It is now open. This method resends the data.
!
! Parameters:
!    time - The time stamp when the button was tapped.

SUB readyToUpdateSubscribers (time AS DOUBLE)
  sendText
END SUB
```

readyToUpdateSubscribers is called if the call to updateValue failed to send the text. Since sendText was careful to leave the unsent text in the line$ variable, all this subroutine has to do is call sendText again.

```
! Handle a tap on a button.
!
! Parameters:
!    ctrl - The button that was tapped.
!    time - The time stamp when the button was tapped.

SUB touchUpInside (ctrl AS Button, time AS DOUBLE)
  IF ctrl = quitButton THEN
    System.showSource
    STOP
  END IF
END SUB
```

The last subroutine is a familiar one. It stops the program when the user taps the Quit button.

This app sets up a single service with two characteristics, but the same technique is used for multiple services and characteristics. It's also possible to add a service to a service to form an included service, or to add descriptors to characteristics. These aren't used often, but if your app needs them, the model followed is so similar to what you've seen in this app that you should have no trouble putting them to use.

Paddles: A Bluetooth Pong Tribute

About This Chapter

Prerequisites

Read Chapter 9 first to get a basic understanding of Bluetooth low energy slave calls.

Equipment

You will need three iOS devices. Each must be an iPhone 4S or later, an iPad 3 or later, or an iPod 5th Gen or later running iOS 6 or later.

Software

You will need a copy of techBASIC or techBASIC Sampler.

What You Will Learn

This project puts what you've learned in the past few chapters to the test as you develop a game that uses iPhones as game paddles. It shows explicitly how to connect to two Bluetooth low energy devices at the same time, as well as how easy it is to set up a Bluetooth low energy slave device.

The Classic Game of Pong

Back in 1974, a contractor at Atari named Steve Jobs teamed up with his friend Steve Wozniak to create a new version of an electronic game called Pong. They were creating a single-player version of this trendy video game. For a whopping $700, plus a bonus for low chip count, the two created the game in four days. Soon they teamed up for another project: building the legendary Apple computer.

It seems only fitting that we return Apple to its very deep roots. Our project for Bluetooth low energy slave mode is a tribute to Pong called Paddles (Figure 10-1) that uses two iPhones or iPads as game paddles and an iPad for the game display. The paddles use the built-in accelerometer to convert tilt into game paddle position, broadcasting changes to the iPad running the Paddles console. The iPad does the rest, tracking the paddles, moving the ball, and keeping score in this modern tribute to four sleepless days and nights in 1974.

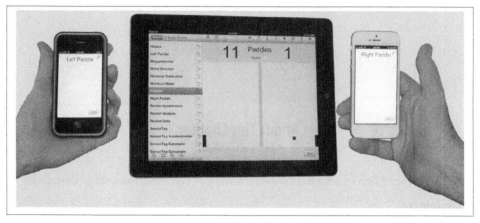

Figure 10-1. The Paddles game

The Paddles Game

Let's take a look at what the program does before trying to understand how it works.

The paddles are really very simple. The GUI consists of a Quit button and a caption telling the users which position they are playing, as shown in Figure 10-2. There are no other indicators—who needs them? After all, it's the iPad people look at while playing the game. The iPad contains the bulk of the user interface.

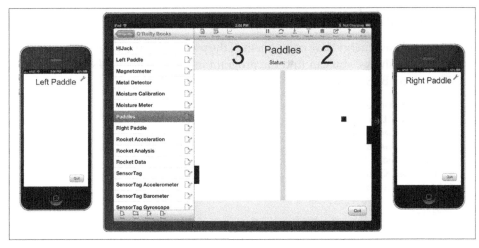

Figure 10-2. The Paddles user interface

As the program starts, it scans for two game paddles. The familiar red-yellow-green status indicator from our earlier applications has changed slightly. It now shows the status for the worst-case connection, displaying red if one paddle has not connected, even if the other is fully functional.

The paddles control the position of the game paddles at the left and right of the screen. As the iPhones tilt, the respective paddles slide to the appropriate location. This happens as soon as the connection is made, even before play starts.

A message appears at the bottom of the screen once both paddles are connected. It's a countdown timer, counting down five seconds so the players are ready when the ball is served. The game ball starts from center court, flinging in a random direction for the first serve, and then shooting toward the player who lost the previous point for subsequent serves. The specific starting position, direction, and speed are random.

It's not enough to just hit the ball. The position of impact on the paddle changes the direction of the ball, too. Hit it in the center, and the ball reflects off of the paddle. Hit it with the edge of a paddle, though, and the direction changes.

Scores are shown at the top of the display. The game ends when one player reaches 15 points.

You now have a choice, and which choice you make reflects a great deal on your character. You can crank up the game now and play a few rounds to get familiar with it, or you can read ahead, learning exactly how the game works to give yourself a competitive advantage before selecting a victim—uh, opponent. It's up to you.

The Paddle Software

The Bluetooth low energy slave component of the project is in the paddle software. Let's start there. As always, you can find the program in the *O'Reilly Books* folder in tech-BASIC and techBASIC Sampler.

```
! This app implements the left
! paddle controller for the
! Paddles game. The Paddles game
! requires two BLE-equipped
! iPhones and a BLE-equipped
! iPad. See Chapter 10 of
! "Building iPhone and iPad
! Electronics Projects" for
! details.

DIM quitButton AS Button
DIM characteristic AS BLEMutableCharacteristic
DIM peripheralManager AS BLEPeripheralManager
DIM lastTime AS Double
orientation = 0
```

The program starts with the familiar declarations for global variables. The Quit button is the only global GUI element, since it's the only one used in multiple subroutines. There is also a variable to hold the BLEMutableCharacteristic so it's not disposed of between subroutine calls, a variable to hold the time the program last updated the Bluetooth low energy data, and a variable recording the screen orientation. We'll see how these are used in a moment.

```
name$ = "LeftPaddle"
serviceUUID$ = "7240D580-B108-11E2-9E96-0800200C9A66"
characteristicUUID$ = "7241D580-B108-11E2-9E96-0800200C9A66"
```

The paddle does one simple thing: it checks the accelerometer to see how far the iPhone is tilted and passes that on to the iPad. Since it only does one thing, the program only needs to create one service. These variables will be used to set up that service and its characteristic.

```
advertise
setUpGUI
```

With the initial variables set up, the program starts advertising so the Paddles game can find the iPhone, and then draws the user interface.

```
! Advertise our presence so the
! Paddles game can find this
! paddle.

SUB advertise
peripheralManager = BLE.newBLEPeripheralManager
```

```
DIM service AS BLEMutableService
service = peripheralManager.newService(serviceUUID$, 1)

characteristic = service.newCharacteristic(characteristicUUID$, $0012, $0001)

peripheralManager.addService(service)
peripheralManager.startAdvertising(name$)
END SUB
```

You saw how to set up a service and advertise it in the last chapter, but the program in that chapter sent and received Bluetooth low energy information. It's easy to miss how easy it is to set up a Bluetooth low energy slave program on the iPhone with both slave and master roles coded in a single program. This subroutine creates the service with a single characteristic, and then advertises it with a specific name. The only other task is to update the name when needed. That's done in the nullEvent subroutine.

```
! Called when the program is not
! busy, this event handler checks
! to make sure 0.25 seconds have
! elapsed. If so, a new paddle
! position is reported to the
! Paddles game.

SUB nullEvent (time AS DOUBLE)
IF time - lastTime > 0.25 THEN
```

nullEvent gets called constantly when the program is not busy. We don't want to clog the airwaves and bog down the iPad with too many updates, so the subroutine starts with a quick check to make sure at least a quarter of a second has elapsed since the last time it posted an update.

Why a quarter of a second? Testing seemed to indicate this was a good value. If the updates come too fast, the Paddles program gets bogged down with the dual task of collecting Bluetooth low energy data from two different sources and updating the graphics screen. These are both time-intensive tasks at the operating system level. It's certainly an area to experiment with. You might find that shorter times will work better in newer versions of iOS or techBASIC.

```
! Get the angle of the device.
a = Sensors.accel
SELECT CASE orientation
  CASE 1 : ! Home button down
    angle = DEG(ANGLE(a(2), a(3)) + PI/2)

  CASE 2 : ! Home button left
    angle = DEG(ANGLE(a(2), a(3)) + PI/2)

  CASE 3 : ! Home button right
    angle = -DEG(ANGLE(a(2), a(3)) + PI/2)
```

```
CASE 4: ! Home button up
   angle = -DEG(ANGLE(a(2), a(3)) + PI/2)
END SELECT
```

Calculations for the angle depend on whether the user is holding the device in portrait or landscape mode, and also vary depending on whether it is upside down. After splitting off to handle the four possibilities, the code to calculate the angle involves a single simple trigonometric calculation. It may look familiar: you saw the same code used in the program that controls the BLE truck.

```
! Send the new angle to the game.
DIM value(1) AS INTEGER
IF angle < -90 THEN
  angle = -90
ELSE IF angle > 90 THEN
  angle = 90
END IF
value(1) = 90 - angle
result% = peripheralManager.updateValue(characteristic, value)
```

We want to send as little data as possible to keep the radio broadcast short and preserve battery life, so the angle will be packed into a single byte. We don't really need angles where the device is rotated so the screen faces down, so the angle is limited to –90 degrees to 90 degrees. That still leaves an issue, though. Bytes are usually thought of as being unsigned, with a valid range of values from 0 to 255. BASIC handles them that way when converting a byte to an integer. We convert the angle from a range of –90 to 90 to a range of 180 to 0 by subtracting the angle from 90, and then blast the new angle out to the game console.

Subtracting the angle from 90 rather than adding 90 may seem like an odd choice. Either will work. The difference is which way the iPhone is tilted to cause the paddle to go up. You can subsitute this code to see which you prefer.:

```
! Send the new angle to the game.
DIM value(1) AS INTEGER
IF angle < -90 THEN
  angle = -90
ELSE IF angle > 90 THEN
  angle = 90
END IF
*value(1) = angle + 90*
result% = peripheralManager.updateValue(characteristic, value)
```

In playability tests, though, the first value seemed to work best.

Thinking back to the SensorTag, you'll now see why the values sometimes had odd conversions to convert the value actually sent from the SensorTag to a number that made sense. Our console will need to convert this value back to an angle by subtracting 90 from the value it gets. The folks who wrote the SensorTag software made many similar decisions.

```
   ! Remember the time so we don't do this constantly.
   lastTime = time
END IF
END SUB
```

The last step is to remember the current timestamp so it's updated for the next call to nullEvent. The remaining two subroutines should look pretty familiar by now:

```
! Set up the user interface.

SUB setUpGUI
! Switch to the graphics screen.
System.showGraphics(1)
System.setAllowedOrientations(1 << (System.orientation - 1))
orientation = System.orientation

DIM title AS Label
title = Graphics.newLabel(20, 20, Graphics.width - 40, 45)
title.setFont("Sans-serif", 40, 0)
title.setText("Left Paddle")
title.setAlignment(2)
title.setBackgroundColor(0, 0, 0, 0)

quitButton = Graphics.newButton(Graphics.width - 92, Graphics.height - 57)
quitButton.setTitle("Quit")
quitButton.setBackgroundColor(1, 1, 1)
quitButton.setGradientColor(0.7, 0.7, 0.7)
END SUB
```

The user interface is very simple, and so is the setUpGUI subroutine. It records the orientation and locks it in so twisting the iPhone during game play doesn't cause the screen to rotate, then creates the label and Quit button.

```
! Handle a tap on one of the buttons.
!
! Parameters:
!    ctrl - The button that was tapped.
!    time - The time when the event occurred.

SUB touchUpInside (ctrl AS Button, time AS DOUBLE)
IF ctrl = quitButton THEN
  STOP
END IF
END SUB
```

touchUpInside stops the program when the user presses the Quit button.

The corresponding program for the right game paddle is almost identical to this one. There is a new name, of course, and the UUIDs for the service and characteristic are different, but other than that, the programs are the same.

The Paddles Console Software

A question I see a lot in programming forums like Stack Overflow and in my own email inbox is how to talk to two Bluetooth low energy devices at the same time. The Paddles game certainly does that—it's communicating with two different Bluetooth low energy game paddles.

Paddles is also the first example of an arcade game in the book. There are some interesting aspects to handling game play to make the game fun and fairly realistic.

With those exceptions, though, there is very little new in this section. If you want to explore setting up two devices or learn more about how to write arcade programs in techBASIC, read on. If not, it's safe to skip this section.

```
! This app implements an electronic version of table tennis, also known as
! ping pong. It requires two BLE-equipped iPhones and a BLE equipped iPad. See
! Chapter 10 of "Building iPhone and iPad Electronics Projects" for details.

DIM quitButton AS Button
DIM leftScoreLabel AS Label, rightScoreLabel AS Label, gameTime AS Label
DIM leftScore, rightScore
DIM leftPaddle AS BLEPeripheral, rightPaddle AS BLEPeripheral
DIM leftPaddleLabel AS Label, rightPaddleLabel AS Label
DIM ballLabel AS Label

leftServiceUUID$ = "7240D580-B108-11E2-9E96-0800200C9A66"
leftCharacteristicUUID$ = "7241D580-B108-11E2-9E96-0800200C9A66"
rightServiceUUID$ = "7242D580-B108-11E2-9E96-0800200C9A66"
rightCharacteristicUUID$ = "7243D580-B108-11E2-9E96-0800200C9A66"
```

Following the now-familiar pattern, the program starts with declarations of global controls and UUIDs.

There are two different sets of UUIDs, one for the left game paddle and one for the right game paddle. What if we wanted to connect to two devices that offered the same service, though, like two SensorTags? That's not difficult, either. While the service and characteristic UUIDs will be the same, the device has a UUID, too. Think of it as the Bluetooth low energy serial number for the device. It's different for each one, and allows the operating system to tell them apart. Our programs can tell them apart the same way.

```
DIM status AS Label
leftStatus = 0
rightStatus = 0

top = 120
bottom = Graphics.height - 80

paddleHeight = 70
paddleWidth = 20
```

```
DIM ballX, ballY, ballVX, ballVY
ballX = -2*ballSize
serveLeft = INT(RND(1)*2)
ballSize = paddleWidth

DIM lastTime AS DOUBLE, gameTimer AS DOUBLE, startGame AS INTEGER
```

A few more variables are defined for drawing the screen and tracking the ball. We'll see how these are used as we explore the rest of the program.

```
scanForPaddles
setUpGUI
```

With the variables initialized, the last step is to start scanning for paddles and set up the user interface.

```
! Begins scanning for paddles.

SUB scanForPaddles
BLE.startBLE
DIM uuid(0) AS STRING
BLE.startScan(uuid)
END SUB
```

The scanForPaddles subroutine starts the discovery process, scanning for both a left and a right game paddle. The Bluetooth low energy subroutines to handle discovery are slight variations on the same stock subroutines you've seen in other Bluetooth low energy programs.

```
! Called when a peripheral is found. If it is a Paddles paddle, we
! initiate a connection to it and stop scanning for peripherals.
!
! Parameters:
!    time - The time when the peripheral was discovered.
!    peripheral - The peripheral that was discovered.
!    services - List of services offered by the device.
!    advertisements - Advertisements (information provided by the
!        device without the need to read a service/characteristic)
!    rssi - Received Signal Strength Indicator

SUB BLEDiscoveredPeripheral (time AS DOUBLE, _
                             peripheral AS BLEPeripheral, _
                             services() AS STRING, _
                             advertisements(,) AS STRING, _
                             rssi)
FOR i = 1 TO UBOUND(advertisements, 1)
  IF advertisements(i, 1) = "kCBAdvDataLocalName" THEN
    IF advertisements(i, 2) = "LeftPaddle" THEN
      BLE.connect(peripheral)
      leftPaddle = peripheral
      leftStatus = 1
      setStatus
    ELSE IF advertisements(i, 2) = "RightPaddle" THEN
```

```
        BLE.connect(peripheral)
        rightPaddle = peripheral
        rightStatus = 1
        setStatus
      END IF
    END IF
  NEXT
  END SUB
```

This program is looking for two peripherals, not one, so it doesn't stop scanning as soon as a Bluetooth low energy peripheral is found. The stopScan call is still important—it cuts battery use when we stop looking for peripherals—but it's been moved to the setStatus subroutine, which we'll look at in a moment. Since stopScan doesn't get called, the operating system keeps right on looking, eventually finding the second paddle.

Whichever paddle is found, the program initiates a connection, saves the peripheral so the connection doesn't get closed by garbage collection on the variable, sets the status for the paddle to 1 to indicate it is in discovery, and calls setStatus to keep the user informed.

```
! Call when the status changes.
!
! Updates the status indicator and, if both paddles are now connected, starts
! a game.

SUB setStatus
print "leftStatus = "; leftStatus; ", rightStatus = "; rightStatus
IF leftStatus = 0 OR rightStatus = 0 THEN
  status.setBackgroundColor(1, 0, 0)
ELSE IF leftStatus = 1 OR rightStatus = 1 THEN
  status.setBackgroundColor(1, 1, 0)
ELSE
  status.setBackgroundColor(0, 1, 0)
  BLE.stopScan
  startGame = 1
END IF
END SUB
```

setStatus is responsible for keeping track of whether the paddles are connected. To keep things simple, there is only one status label. This is a design decision. The program could have had two status indicators, one for each paddle, but I thought that might be confusing for the user. Maybe you disagree. Maybe you're one of *those* people, who want a separate indicator for each paddle. You have the source. Go for it.

This subroutine checks to see if the status of either paddle is 0, painting the status label red if so. If both are greater than 0, it drops through to the next check, seeing if either paddle's status is 1, indicating it has been discovered but no connection has been made. In that case, the status label is painted yellow.

If the status for both paddles is 2, indicating a full connection for both paddles, the subroutine drops through and does three things. Of course, it paints the status green, but it also stops scanning for new peripherals. The last step is to start the game. We'll see how the startGame variable is used later.

That's really all there is to connecting to two different Bluetooth low energy devices at the same time. You could, of course, create an array of peripherals and track even more than two. The rest of the program would need to be adjusted slightly to handle more than one peripheral variable, and you'd have to decide how to make use of all that information and display it to the user in a reasonable way, but that's pretty straightforward coding.

Getting back to the code, let's take a look at the BLEPeripheralInfo subroutine:

```
! Called to report information about the connection status of the
! peripheral or to report that services have been discovered.
!
! Parameters:
!    time - The time when the information was received.
!    peripheral - The peripheral.
!    kind - The kind of call. One of
!        1 - Connection completed
!        2 - Connection failed
!        3 - Connection lost
!        4 - Services discovered
!    message - For errors, a human-readable error message.
!    err - If there was an error, the Apple error number. If there
!        was no error, this value is 0.

SUB BLEPeripheralInfo (time AS DOUBLE, _
                       peripheral AS BLEPeripheral, _
                       kind AS INTEGER, _
                       message AS STRING, _
                       err AS LONG)
DIM uuid(0) AS STRING
SELECT CASE kind
  CASE 1 ❶
    peripheral.discoverServices(uuid)

  CASE 2, 3 ❷
    IF leftPaddle <> NULL AND peripheral.uuid = leftPaddle.uuid THEN
      leftStatus = 0
      BLE.connect(leftPaddle)
    ELSE IF rightPaddle <> NULL AND peripheral.uuid = rightPaddle.UUID THEN
      rightStatus = 0
      BLE.connect(rightPaddle)
    END IF
    setStatus

  CASE 4 ❸
    DIM services(1) AS BLEService, included(1) AS BLEService
```

```
        services = peripheral.services
        FOR i = 1 TO UBOUND(services, 1)
          IF services(i).uuid = leftServiceUUID$ THEN
            peripheral.discoverCharacteristics(uuid, services(i))
          ELSE IF services(i).uuid = rightServiceUUID$ THEN
            peripheral.discoverCharacteristics(uuid, services(i))
          END IF
        NEXT
    END SELECT
    END SUB
```

Here's what the subroutine is doing:

❶   BLEPeripheralInfo is called the first time with a kind of 1, when a connection is established. We then start the process of discovering services.

❷   kind will be 2 or 3 if the connection is lost. The program checks to see which peripheral lost the connection and attempts to reconnect. It also changes the status back to yellow. This is pretty crucial for a game. As you will see in a moment, the game will pause if the status of both paddles is not 2, giving the player who dropped the connection a bit of a chance to recover.

❸   Once the connection is established, iOS calls the BLEPeripheralInfo subroutine with kind set to 4. The program starts discovery for the characteristic. The characteristic varies for the left and right paddles, so it's important to keep track of which paddle made the connection.

Once discovery is complete for a characteristic, BLEServiceInfo tells the paddle to notify this program when the value has been updated. Other than tracking two paddles, this is pretty standard stuff by now:

```
! Called to report information about a characteristic or included
! services for a service. If it is one we are interested in, start
! handling it.
!
! Parameters:
!    time - The time when the information was received.
!    peripheral - The peripheral.
!    service - The service whose characteristic or included
!        service was found.
!    kind - The kind of call. One of
!        1 - Characteristics found
!        2 - Included services found
!    message - For errors, a human-readable error message.
!    err - If there was an error, the Apple error number. If there
!        was no error, this value is 0.

SUB BLEServiceInfo (time AS DOUBLE, _
                    peripheral AS BLEPeripheral, _
                    service AS BLEService, _
                    kind AS INTEGER, _
```

```
                    message AS STRING, _
                    err AS LONG)
  IF kind = 1 THEN
    DIM characteristics(1) AS BLECharacteristic
    characteristics = service.characteristics
    FOR i = 1 TO UBOUND(characteristics, 1)
      IF characteristics(i).uuid = leftCharacteristicUUID$ THEN
        peripheral.setNotify(characteristics(i), 1)
        leftStatus = 2
        setStatus
      ELSE IF characteristics(i).uuid = rightCharacteristicUUID$ THEN
        peripheral.setNotify(characteristics(i), 1)
        rightStatus = 2
        setStatus
      END IF
    NEXT
  END IF
END SUB
```

BLECharacteristicInfo is called when the angle of a paddle is updated. It checks to see which paddle moved and updates it:

```
! Called to return information from a characteristic.
!
! Parameters:
!    time - The time when the information was received.
!    peripheral - The peripheral.
!    characteristic - The characteristic whose information
!        changed.
!    kind - The kind of call. One of
!        1 - Called after a discoverDescriptors call.
!        2 - Called after a readCharacteristics call.
!        3 - Called to report status after a writeCharacteristics
!            call.
!    message - For errors, a human-readable error message.
!    err - If there was an error, the Apple error number. If there
!        was no error, this value is 0.
!
SUB BLECharacteristicInfo (time AS DOUBLE, _
                           peripheral AS BLEPeripheral, _
                           characteristic AS BLECharacteristic, _
                           kind AS INTEGER, _
                           message AS STRING, _
                           err AS LONG)
  IF kind = 2 THEN
    DIM value(1) AS INTEGER
    value = characteristic.value
    IF characteristic.UUID = leftCharacteristicUUID$ THEN
      movePaddle(1, value(1))
    ELSE IF characteristic.UUID = rightCharacteristicUUID$ THEN
      movePaddle(0, value(1))
    END IF
```

```
    ELSE IF kind = 3 AND err <> 0 THEN
      PRINT "Error writing "; characteristic.uuid; ": ("; err; ") "; message
    END IF
    END SUB
```

movePaddle does a bit of math to transform the angle to a paddle position and updates the location of the paddle:

```
    ! Move a paddle.
    !
    ! Parameters:
    !    isLeft - True to move the left paddle, or false to move the right paddle.
    !    angle - The angle at which the paddle controller is held.

    SUB movePaddle (isLeft AS INTEGER, angle)
    tilt = 40
    angle = angle - 90
    IF angle < -tilt THEN
      angle = -tilt
    ELSE IF angle > tilt THEN
      angle = tilt
    END IF
      space = bottom - top - paddleHeight
      y = top + space*(angle + tilt)/(2*tilt)
    IF isLeft THEN
      leftPaddleLabel.setFrame(0, y, paddleWidth, paddleHeight)
    ELSE
      rightPaddleLabel.setFrame(Graphics.width - paddleWidth, y_,
      paddleWidth, paddleHeight)
    END IF
    END SUB
```

This is one place where the program could be improved a bit. I decided to keep it simple so the program was easier to understand, moving the paddle to the new position immediately. From a playability standpoint, though, it might be better if the paddles didn't jump to the new position immediately, but moved at a maximum velocity toward the new position. This would force the players to think ahead a bit more.

The paddles can be moved even before the game starts. The position is checked during play, but the process of moving the paddles is handled here, completely separate from the logic that tracks the ball and keeps score.

It probably comes as no surprise that the active logic to control the game is in nullEvent:

```
    ! Called when the program is not doing anything else, this subroutine
    ! moves the ball every 0.1 seconds. It also handles the start of game
    ! timer if a game has not yet started.
    !
    ! Parameters:
    !    time - The time when the call was made.

    SUB nullEvent (time AS DOUBLE)
```

```
IF time - lastTime > 0.1 AND leftStatus = 2 AND rightStatus = 2 THEN
  moveBall
  lastTime = time
END IF
```

nullEvent starts with the familiar check to make sure some time has elapsed since the last call, then moves the ball. Most of the game logic is implemented in moveBall. That's where the program decides if the ball has hit something and needs to bounce, and if so, which direction it should take. If the ball whips past a paddle, the score is updated and a new ball is served. In true top-down program design, we won't worry about those details until we look at moveBall.

Back in setStatus, the program set startGame to 1 when the status of both paddles reached a value of 2. Here's where that value is used to actually start the game:

```
IF startGame AND leftStatus = 2 AND rightStatus = 2 THEN
  startGame = 0
  gameTimer = time + 5
  ballLabel.setHidden(1)
  ballX = Graphics.width/2
  ballVX = 0
END IF
```

The status is checked again because setStatus is not the only place a game can be started. moveBall will also start a game if one player's score hits 15.

The game doesn't start right away, though. As you know from playing the game, there is a five-second countdown timer so the players have time to get their game faces on. Paddles can be very competitive. You need time to prepare.

The program also hides the ball and sets its position and velocity. That might seem redundant, but it simplifies the game logic in moveBall quite a bit. This code stops the ball so moveBall won't get fooled and track an invisible ball, continuing to rack up points for bewildered opponents.

The last bit of code handles the countdown timer:

```
IF gameTimer > 0 THEN
  IF gameTimer < time THEN
    gameTimer = 0
    gameTime.setText("")
    leftScoreLabel.setText("0")
    leftScore = 0
    rightScoreLabel.setText("0")
    rightScore = 0
    serve
  ELSE
    t = 1 + INT(gameTimer - time)
    gameTime.setText("Game starts in: " & STR(t))
  END IF
```

```
  END IF
END SUB
```

When the time hits zero, both scores are reset to 0 and `serve` is called to, ahem, start the ball rolling:

```
! Serve the ball.
!
! Call this subroutine when it is time to serve the ball.

SUB serve
ballX = Graphics.width/2 ❶
ballY = top + (bottom - top - ballSize)*RND(1)
ballVX = 15 + 10*RND(1) ❷
IF serveLeft THEN ballVX = -ballVX
ballVY = 15*(RND(1) - 0.5)
ballLabel.setFrame(ballX, ballY, ballSize, ballSize) ❸
ballLabel.setHidden(0)
END SUB
```

Let's step through this subroutine to see how it works:

❶   `serve` starts by moving `ballX` to center court and `ballY` to a random location so the ball doesn't start in a predictable place.

❷   `ballVX` and `ballVY` track the speed for the ball in the x and y directions. Both are randomized. The horizontal velocity is randomized just a bit; most players won't even notice this slight variation, but it keeps the game a little less predictable. Randomizing the velocity in the y direction changes the initial direction of the ball, again to keep things less predictable.

❸   Like the `status`, the ball is actually a `Label`, so the program sets the position of the ball by changing the location of the label. Since it is hidden while the countdown timer runs, the subroutine ends by making the ball visible.

Most of the game logic is in `moveBall`:

```
! Move the ball.
!
! Moves the ball based on the current ball velocity, then checks for bounces,
! paddle hits, and paddle misses, handling them as appropriate.

SUB moveBall
ballY = ballY + ballVY
```

The first step in moving the ball is to update the position of the ball. The program splits movement along the two axes, starting with y. Using a basic equation from physics, the new position for the ball is:

$$d_1 = d_0 + v \cdot t$$

where d_0 is the original position of the ball, v the velocity, and t the time step. The new position is d_1. By conveniently defining the time step to be 1, all we have to do to update the position of the ball is add the velocity.

Of course, the ball might hit the top or bottom of the court, so the next step is to check for that possibility. When it happens, the program reverses the direction of the ball in the y direction:

```
IF ballY < top THEN
   ballY = top + (top - ballY)
   ballVY = -ballVY
ELSE IF ballY > bottom - ballSize THEN
   temp = bottom - ballSize
   ballY = temp - (ballY - temp)
   ballVY = -ballVY
END IF
```

But the ball might have traveled past the wall by a few pixels, and that would be a jarring departure from the physical world. The program handles that possibility by moving the ball back in the other direction. If the new position was one pixel past the wall, the ball is placed one pixel away from the wall in the other direction. This maintains the illusion of a physical ball bounce by matching what would happen in real life.

Tracking the ball in the x direction is a bit more complicated, since the left and right ends of the court are open. The first step is to see if the ball hit a paddle:

```
ballX = ballX + ballVX
IF ballX < paddleWidth THEN
   ballX = paddleWidth + (paddleWidth - ballX)
   ballVX = -ballVX
   y = leftPaddleLabel.y
   IF ballY < y - ballSize OR ballY > y + paddleHeight THEN
```

The program handles a miss first. If the ball missed the paddle, it's immediately hidden and the score is updated:

```
      ballLabel.setHidden(1)
      rightScore = rightScore + 1
      rightScoreLabel.setText(STR(rightScore))
```

The game is over if the score is 15. That's handled by starting a new game:

```
      IF rightScore = 15 THEN
         startGame = 1
```

This is all in an effort to keep you healthy. You're playing a video game, after all. You have time to go to the refrigerator for more junk food if the game waits too long. The program will only wait for five seconds, so you either have to forgo the junk food or hurry, and thus get some exercise. (You're welcome. It's just one of the small public services programmers perform every day.)

If the score has not hit 15, the program pauses for a second and serves the ball to the player who missed:

```
ELSE
  System.wait(1)
  serveLeft = 1
  serve
END IF
```

There's a lot of game play in this next little section of code. Unlike bouncing off of the top or bottom wall, the ball doesn't necessarily bounce off of the paddle by simple reflection. The *y* velocity is adjusted based on where the ball hit the paddle. A hit near the top of the paddle adds a bit to the velocity in the up direction, while hitting the paddle near the bottom adds some velocity in the down direction.

It's still not a constant, though. The new velocity is added to the old one. A skilled player can hit a ball traveling fast in one direction or another so it speeds up, bouncing at a higher and higher vertical speed, making it more and more difficult for the opponent to hit the ball:

```
ELSE
  dy = 15*((ballY - (y - ballSize))/(ballSize + paddleHeight) - 0.5)
  ballVY = ballVY + dy
END IF
```

The next step is to repeat all of that logic for a ball hitting the right side of the court:

```
ELSE IF ballX > Graphics.width - paddleWidth - ballSize THEN
  temp = Graphics.width - paddleWidth - ballSize
  ballX = temp - (ballX - temp)
  ballVX = -ballVX
  y = rightPaddleLabel.y
  IF ballY < y - ballSize OR ballY > y + paddleHeight THEN
    ballLabel.setHidden(1)
    leftScore = leftScore + 1
    leftScoreLabel.setText(STR(leftScore))
    IF leftScore = 15 THEN
      startGame = 1
    ELSE
      System.wait(1)
      serveLeft = 0
      serve
    END IF
  ELSE
    dy = 15*((ballY - (y - ballSize))/(ballSize + paddleHeight) - 0.5)
    ballVY = ballVY + dy
  END IF
END IF
```

Finally, the ball is moved to its new location:

```
ballLabel.setFrame(ballX, ballY, ballSize, ballSize)
END SUB
```

```
! Set up the user interface.

SUB setUpGUI
DIM title AS Label
title = Graphics.newLabel(20, 20, Graphics.width - 40, 45)
title.setFont("Sans-serif", 40, 0)
title.setText("Paddles")
title.setAlignment(2)
title.setBackgroundColor(0, 0, 0, 0)

! Add a status indicator. The status label is set to red initially,
! turns yellow when a connection is made to both paddles, and
! changes to green when both paddles are started.
DIM statusLabel AS Label
x = Graphics.width/2 - 90
y = 80
width = Graphics.width
statusLabel = Graphics.newLabel(x, y, 100)
statusLabel.setText("Status:")
statusLabel.setBackgroundColor(0, 0, 0, 0)
statusLabel.setAlignment(3)
statusLabel.setFont("Arial", 20, 0)

status = Graphics.newLabel(x + 110, y, 21)
setStatus

! Add the game score indicators, initializing them to 0.
leftScoreLabel = Graphics.newLabel(Graphics.width/8, 10, _
                                   Graphics.width/4, 105)
leftScoreLabel.setFont("Sans-serif", 100, 0)
leftScoreLabel.setText("0")
leftScoreLabel.setAlignment(2)
leftScoreLabel.setBackgroundColor(0, 0, 0, 0)

rightScoreLabel = Graphics.newLabel(Graphics.width*5/8, 10, _
                                    Graphics.width/4, 105)
rightScoreLabel.setFont("Sans-serif", 100, 0)
rightScoreLabel.setText("0")
rightScoreLabel.setAlignment(2)
rightScoreLabel.setBackgroundColor(0, 0, 0, 0)

! Add the game time counter.
width = 200
gameTime = Graphics.newLabel((Graphics.width - width)/2, Graphics.height - 50, width, 30)
gameTime.setFont("Sans-serif", 24, 0)
gameTime.setBackgroundColor(0, 0, 0, 0)

! Fill the non-playing field with gray.
Graphics.setColor(0.9, 0.9, 0.9)
Graphics.fillRect(0, 0, Graphics.width, top)
Graphics.fillRect(0, bottom, Graphics.width, Graphics.height - bottom)
```

```
! Draw the net at center court.
Graphics.setColor(0.8, 0.8, 0.8)
Graphics.fillRect((Graphics.width - paddleWidth)/2, top, _
                  paddleWidth, bottom - top)

! Add the paddles.
leftPaddleLabel = Graphics.newLabel(0, (Graphics.height - paddleHeight)/2, _
                                    paddleWidth, paddleHeight)
leftPaddleLabel.setBackgroundColor(0, 0, 0)

rightPaddleLabel = Graphics.newLabel(Graphics.width - paddleWidth, _
                                     (Graphics.height - paddleHeight)/2, _
                                     paddleWidth, paddleHeight)
rightPaddleLabel.setBackgroundColor(0, 0, 0)

! Add the ball.
ballX = Graphics.width/2
ballVX = 0
ballLabel = Graphics.newLabel(ballX, 0, ballSize, ballSize)
ballLabel.setHidden(1)
ballLabel.setBackgroundColor(0, 0, 0)

! Add the Quit button.
quitButton = Graphics.newButton(Graphics.width - 92, Graphics.height - 57)
quitButton.setTitle("Quit")
quitButton.setBackgroundColor(1, 1, 1)
quitButton.setGradientColor(0.7, 0.7, 0.7)

System.showGraphics
END SUB
```

There are a number of controls, text fields, and so forth on the screen, so setUpGUI is a little longer than in some programs, but it's all stuff you've seen before. The same is true for touchUpInside, which handles the Quit button:

```
! Handle a tap on a button.
!
! Parameters:
!    ctrl - The button that was tapped.
!    time - The time stamp when the button was tapped.

SUB touchUpInside (ctrl AS Button, time AS DOUBLE)
IF ctrl = quitButton THEN
  STOP
END IF
END SUB
```

So there you have it. Our simple game is a lot of fun, but it also shows how to connect multiple Bluetooth low energy slave devices to a single master device.

CHAPTER 11

WiFi

About This Chapter

Prerequisites

Read Chapter 1 and the end of Chapter 2 (the section on the techBASIC help system) if you need some help with techBASIC.

Equipment

You will need an iPhone, iPod, or iPad running iOS 5 or later. You will also need a WiFly device, an Arduino Uno, and various common electronics parts like a small regulated power supply, jumper wires, and a breadboard. See Table 11-1 for a complete parts list.

Software

You will need a copy of techBASIC or techBASIC Sampler and a copy of the Arduino software for programming the Arduino. The Arduino software is a free download.

What You Will Learn

This project shows how to use a WiFi to serial bridge to set up communication with practically any device that uses serial input.

Worldwide Sensors

Bluetooth low energy is a great way to communicate over short distances, but it does have a couple of limitations. The first is *bandwidth*, or the amount of information it can move in a given amount of time. Bluetooth low energy typically delivers 100 Kbps (kilobits per second) or less. The second is *range*. Depending on the antenna and what the radio waves need to pass through, Bluetooth low energy is limited to a range of about 20 to 100 meters.

Our next technology is a bit better in both areas. WiFi connects your device to the Internet, where, with the proper passwords, you can communicate with the International Space Station. Just a tad further than 100 meters, right? As for bandwidth, it depends a lot on what you are connected to, but typical speeds are 20–35 Mbps (megabits per second) for download, and 2–10 Mbps for upload.

WiFi is actually a radio technology for networking. It replaces the common Ethernet cable used to connect desktop computers and other electronics that stay in a fixed location with a radio signal. The standard for WiFi communications is IEEE 802.11, which defines all of the speeds, frequencies, and protocols used for wireless networking. That's why you frequently see stuff in advertisements for wireless routers that talk about them being 802.11, generally with some sort of letter indicating a subportion of the standard. The iPhone and some iPads supplement normal WiFi with cellular phone connections. The operating system takes care of most of the details—for the most part, our programs just need to know the Internet address of the thing we want to talk to, and iOS handles all the details of making the connection with the best method available.

HTTP, FTP, and TCP/IP

There are a lot of different ways to communicate with devices using networks. It helps to know just a little about network communications to understand what each is good for. This explanation is simplified a bit—entire books have been written on a topic we're going to cover in a few paragraphs—but it should give you enough information to understand what we'll be doing later.

Modern computers use an addressing scheme built on an *IP* (Internet Protocol) address. This is really a four-byte value, but it's invariably written as four numbers that range from 0 to 255, separated by dots. For example, 173.194.46.8 is an IP address you probably use almost every day. Most folks are not particularly good at remembering numbers, though, which is why the *domain name* was invented. A domain name is simply a name for an IP address. There are computers on the Internet whose sole reason for existing is to keep track of the names associated with IP addresses, so when you type *http://www.google.com* into your browser, the computer knows it wants to talk to 173.194.46.8.

Discussions between computers occur on *ports*. Two bytes are used to identify the ports, giving 65,536 total ports numbered 0 to 65535. Some ports have predefined uses, while others are open for pretty much any use imaginable. By far the most commonly used port is port 80, which is used for browser communication. That, combined with a pre-arranged format for the text passed over port 80, gives us what most people think of as the Internet—the familiar *HTTP* (Hypertext Transfer Protocol) used by browsers to display and collect information.

Probably the second most common use of the Internet is to move files back and forth from one computer to another. This is often handled in ways that are invisible to us, taking place behind the scenes in the browser. We can, however, make use of port 21.

This port is the default port for *FTP* (File Transfer Protocol). Using an FTP client, this protocol lets you treat a hard drive on another computer almost as if it were physically installed on your own computer. You can see files, rename them, delete or create files, or move files to and from the other computer using FTP. Of course, you generally need an account and a password to do this.

But what if you want to do something unusual? There are a couple of common technologies for communicating directly with a port, sending and receiving either text or raw bytes. The one we'll use in this chapter is *TCP/IP* (Transmission Control Protocol/ Internet Protocol). It's designed for two-way communications between two computers using an agreed-upon port, which can be any of the 65,536 available ports. This is one of the basic ways of communicating between two computers. It's perfectly possible to write your own FTP client or browser using just TCP/IP, and for the most part, that's probably what's happening behind the scenes in the browsers and FTP clients you use every day.

techBASIC has support for all of these protocols, but we'll only be using TCP/IP. You can explore the others in the Comm class if you are curious.

WiFly

Our program will need something to talk to on the other end. The number of possible devices you can control using TCP/IP is staggering, from telescopes to traffic lights. In our case, we'll use a versatile little device called the *WiFly*, or more formally, the Roving Networks RN-XV (Figure 11-1). There are several variants of this device, mostly offering different ways to mount the device or different antenna configurations. The one we'll use is designed to plug into mounts for the popular ZigBee devices. It uses a simple wire antenna mounted right on the device.

Figure 11-1. The Roving Networks RN-XV WiFly

The WiFly essentially consists of a tiny computer, a wireless networking router, some programmable I/O units, and a serial port mounted on a single device that can be powered with a coin cell battery. There are lots of things you can do with the device, but we'll concentrate on using it to convert from WiFi to serial, allowing us to communicate with any serial device from the iPhone. Everything you need to know for the projects in this book is covered here. To learn more, start with the User Manual and Command Reference, which you can download from the Roving Networks site (*http://www.roving networks.com/files/resources/WiFly-RN-UM.pdf*). The pinouts are in the data sheet, which can be downloaded from the Resources and Documentation section of the website (*http://www.rovingnetworks.com/resources/download/16/RN_XV*).

Table 11-1 lists the parts we'll use, along with some alternatives.

Table 11-1. Parts list

Part	Description
Any iPhone, iPad, or iPod	They all have WiFi; another advantage over Bluetooth low energy.
techBASIC	We'll use the WiFly Terminal program, preinstalled in techBASIC. You will need to change an IP address, so techBASIC Sampler will only work if you use the in-app purchase to get the editing capabilities.
WiFly	More formally, the RN-XV WiFly Module with Wire Antenna, or WRL-10822. You can use other mountings or antennas if you like.
XBee Breakout	This converts the 20-pin 0.2 mm spacing of the standard XBee device to the 0.1" pin spacing used on breadboards. It's a handy way to prototype an application. These are available from several sources, including SparkFun (*http://www.sparkfun.com*), where the part number is BOB-08276.
2 mm 10-pin XBee Header (2)	You will need two 1 × 10 headers with a 2 mm spacing for the XBee Breakout board. These are an unusual spacing, so plan ahead—you can't pick these up at most local electronics stores. These are also available from SparkFun.
Break Away Headers - Straight	You will also need two 1 × 10 headers with 0.1" spacing for the XBee Breakout board. These are the common headers you can find almost anywhere. This specific part is from SparkFun, and is a 1 × 40 header you can clip apart to any desired length.
DFRobot Breadboard Power Supply 5V/ 3.3V	I used this versatile power supply, which can grab power from a USB port or from a standard DC power supply with a 2.1 central positive jack. You can use any power supply that provides a steady 3.3 volts and 5 volts, though, from a battery to a home-built regulated power supply.
Breadboard and jumpers	You'll also need a breadboard and some jumper wires for connecting everything.
Arduino Uno	You can do the basics without the Arduino—it's really used as a practical way to put the connection to use. You can use any Arduino that has serial communications; the Arduino Uno is simply the most common and widely available of these.

The Circuit

We'll be using a *null modem connection* with the serial I/O to set up and test the WiFly. This means it will be hooked up so any text we send from our iOS device is echoed back to us. If everything works properly, the device will become a perfect mimic.

The circuit itself is very simple, as shown in Figure 11-2. Power is supplied by connecting pin 1 (the one at the top left of the device) to +3.3V, and pin 10 (bottom left) to ground.

We also connect pin 8 to +3.3V; this tells the device to default to Ad Hoc mode, where it sets up its own network instead of connecting to an existing one. It becomes its own little network island, and will even show up on your desktop computer as a WiFi hotspot.

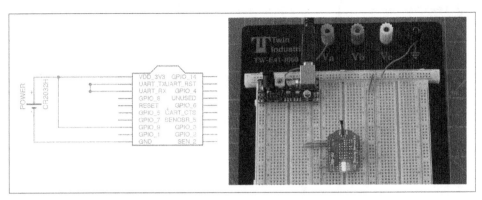

Figure 11-2. The null modem circuit

Pins 2 and 3 handle serial I/O. Pin 2 is UART_TX, which is the output pin for serial I/O. When we send a character from the iPhone, it will show up on this pin as a series of high and low voltage levels. Pin 3 is UART_RX, which watches for high and low voltage levels, combining these into bytes that it sends back to the iOS device. We'll wire these together so anything we send out comes right back to the program.

Establishing a Network Connection

Once everything is wired, apply power. You should see three blinking LEDs on the device; one red, one green, and one yellow. The yellow one indicates whether anything has established a connection with the WiFi hotspot the device set up.

Go to the Settings app on your iPhone or iPad and select Wi-Fi, right near the top. In a moment your list of network devices will expand. One of the networks will be *Wifly-GSX-xx*, where *xx* will be a two-digit hexadecimal value, as shown in Figure 11-3. This changes for different devices; for mine, the value was *d3*. Tap this line to select your WiFly device. After a moment, your WiFly network will be selected. The yellow LED will also stop flashing on the WiFly device itself.

Figure 11-3. Connecting to the WiFly

Tap the blue disclosure button to the right of the device to see more information about the connection. One item will be the IP address for the network; mine was 169.254.46.56, but the last two numbers varied, even with different connections to the same device.

> Always wait for the IP address to show up here before trying to make a connection. The connection may not be completely established until the address appears. If you disconnect and reconnect, double-check the IP address before trying to connect from techBASIC.

Your iPhone is now connected to the WiFly module. Programs can talk to it using TCP/IP.

It's worth pointing out that there are a lot of different ways to connect your iPhone or iPad to the WiFly device. This method is simple and works really well for dealing with sensors or other devices connected to a single WiFly. It also doesn't require any other outside hardware. It's possible to connect the WiFly to your existing WiFi network, which is a great option if you have one and the WiFly sensor you are building will always be used within range of your existing WiFi network. Check the documentation for the device to determine how to set up different kinds of connections, modify the connection settings, and so forth.

Communication with TCP/IP

Communicating with a WiFly from techBASIC is deceptively simple. You open a file as a TCP/IP connection, then use standard BASIC I/O commands to talk to it.

The command to open the connection is:

```
Comm.openTCPIP(1, "169.254.1.1", 2000)
```

The first parameter is the channel number, used in the various BASIC I/O commands to identify which file to use. Next is the IP address. The last two numbers are always 1.1 for a WiFly connected in Ad Hoc mode; the first two numbers match the address of the network. The last value is the port number, which is always 2000 for the WiFly.

Once a communications channel is open, standard I/O commands are used. For example, to send the command "Hello, WiFly" to the device, use:

```
PRINT #1, "Hello, WiFly"
```

The PUT command can be used if you are sending binary values rather than characters.

Getting information back is not hard, but you have to keep in mind that the device may not have anything to send. The easiest way to check to see if the device has anything to send back is to check EOF. If it returns false, there is information available; if it returns true, there is nothing to read. Since our program will probably want to do other things if the device is not sending information back, the best place to check to see if anything is available is in a nullEvent subroutine. We've seen those before—they are called when the program is not busy doing something else, like handling a button tap. We know the WiFly is set up to send back whatever we send to it, so the obvious way to read information once some is available is to use a LINE INPUT command, which will grab one line of text and return it all in one string. Unfortunately, there is a practical problem with using LINE INPUT. The device will send other information back, and not all of it ends with an end-of-line mark. To get around this little issue, we'll grab information one byte at a time using the GET command, processing it in our program as appropriate.

Other BASIC I/O commands work with TCP/IP streams, too, but these are the only ones we need for the program we'll use to send and receive data to and from the WiFly device. See the techBASIC Reference Manual (*http://www.byteworks.us/Byte_Works/Documentation.html*) or built-in help, especially the section on Comm.openTCPIP, for information about the other I/O commands.

A Simple Terminal Program

The basic software to communicate with the WiFly is a pretty simple text-based terminal program. We'll get fancy in the next chapter.

```
PRINT "Simple WiFly Terminal"
PRINT
```

```
PRINT "Requires a Roving Networks WiFly module to function."
PRINT
Comm.openTCPIP(1, "169.254.1.1", 2000) ❶
DIM t AS DOUBLE ❷
t = System.ticks

SUB nullEvent (time AS DOUBLE)
IF System.ticks - t > 0.25 THEN ❸
  PRINT
  LINE INPUT "> "; a$ ❹
  PRINT #1, a$ ❺
  t = System.ticks ❻
ELSE
  WHILE NOT EOF(1) ❼
    GET #1,,b~ ❽
    PRINT CHR(b~); ❾
  WEND
END IF
END SUB
```

Let's take a look at what the program is doing here:

❶   After printing a header to remind us what program we're running, the program opens the TCP/IP port. Change this line to match the IP address for your WiFly device.

❷   The program will let the user type a command, send the command to the WiFly device, and then wait a quarter of a second for a response. It will print anything sent back from the device during that time, then let the user type a new command. It's a very simply way to communicate, but it works remarkably well for learning about the device. These lines set up a value used as a timer and grab an initial time.

❸   The nullEvent subroutine is called continuously. It starts with a check to see if a quarter of a second has elapsed. If so…

❹   …the program stops and asks the user for a command.

❺   Next it sends the command to the WiFly device.

❻   We then reset the timer so we will wait at least a quarter of a second before waiting for more input. We need to do this so we don't miss values returned from the device while the LINE INPUT command blocks other program execution, waiting for the user to type something.

❼   The program then checks to see if there is anything to read. If not, the subroutine will exit right away, get called again, and we'll start the checks again.

❽   This line gets one character from the device.

❾   Finally, we print the character to the console.

The program is called WiFly Terminal. It's in the *O'Reilly Books* folder in techBASIC and techBASIC Sampler. Run it and the program will print the header lines, then something you may not have expected:

```
Simple WiFly Terminal

Requires a Roving Networks WiFly module to function.

*HELLO**OPEN*
```

The extra line at the end is an initial status message sent by the WiFly device itself. It does not end with an end-of-line mark, which is why we needed to use GET rather than LINE INPUT. Don't worry if it does not show up.

Type "Hello, WiFly." at the prompt. The device should echo the string back:

```
> Hello, WiFly.
Hello, WiFly.
```

WiFi Arduino

The next modification to our project replaces the null modem connection with an Arduino microcontroller. It's an amazingly simple change.

Loading Software onto the Arduino

The first step is actually the hardest, though it's pretty routine for Arduino fans. We need to place a program on the Arduino that will do something when we send a command over the serial port.

 This section assumes you are familiar with installing and running the Arduino software. If this is your first time, refer back to "Installing Arduino" on page 219 for detailed instructions on setting up the software for the first time.

Start by connecting the Arduino Uno to your desktop computer using a USB cable. Run the Arduino app used to upload programs to the Arduino (available from the Arduino website (*http://arduino.cc/en/main/software*)). Once the Arduino app is running, pull down the Tools menu and check the Board setting to make sure the exact Arduino model you are using is selected, as shown in Figure 11-4.

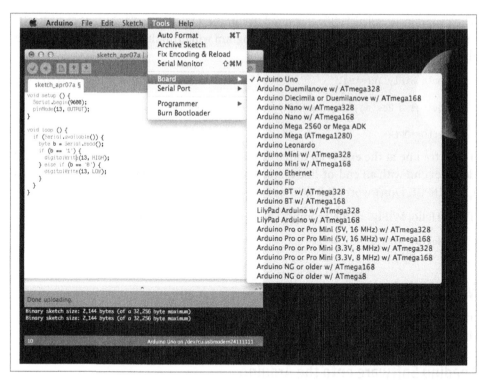

Figure 11-4. Make sure your Arduino model is selected

Next, check the Serial Port setting to make sure the correct serial port is selected, as shown in Figure 11-5.

The program we'll use turns on an LED on the Arduino when a 1 is sent over the serial connection, and turns it off when a 0 is sent. Pull down the File menu, select New, and enter this program into the Sketch window:

```
void setup () {
  Serial.begin(9600);
  pinMode(13, OUTPUT);
}

void loop () {
  if (Serial.available()) {
    byte b = Serial.read();
    if (b == '1') {
      digitalWrite(13, HIGH);
    } else if (b == '0') {
      digitalWrite(13, LOW);
    }
  }
}
```

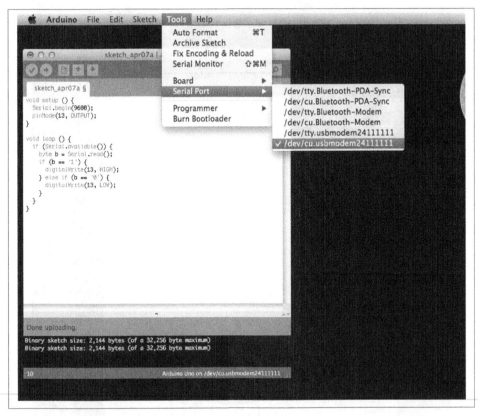

Figure 11-5. Make sure the serial port is correct

The program opens a serial connection, sets pin 13 to be an output pin, and then drops into a loop waiting for something to happen. This loop is the equivalent of the nullE vent subroutine in techBASIC. When a character is detected, the program checks to see if it is a 0 or 1, ignoring anything else. If the character is a 0 or 1, the LED is turned off or on.

The Circuit

Now that it is programmed to do your bidding, disconnect the Arduino from the desktop computer and connect the pin labeled 5V to the positive side of the 5V connection on the power supply, and one of the pins labeled GND to the negative side of the 5V connection on the power supply. (You can see the circuit in Figure 11-6.) This provides power. You can also provide power by leaving the Arduino board plugged into the desktop computer; it will draw power from the USB port.

Figure 11-6. Arduino/WiFly circuit

Connect the pin labeled TX-1 to pin 3 on the WiFly and the pin labeled RX-0 to pin 2 on the WiFly. If you recall from earlier, pin 3 is the RX pin on the WiFly, and pin 2 is the TX pin. We've just wired the two boards so anything transmitted over the serial port to the WiFly is passed on as input to the Arduino, and anything the Arduino sends back is sent to the receive pin on the WiFly, where it will be passed back to the iOS device.

Communication Using the Terminal Program

Crank up the same WiFly Terminal program you've already used and type a **1**. The LED on the Arduino will light up as shown in Figure 11-7. Now type a **0**. It turns off.

Figure 11-7. The LED when it is lit

OK, maybe that's not as exciting as, say, operating a robot, but guess what? You now have all of the hardware and software you need to do just that. Entire books are devoted to doing little more than getting LEDs to light on an Arduino. Why? Because it's a general-purpose microcontroller, and whole books, countless Internet blogs, and lots of magazine articles have also been written on different things you can do with an Arduino—like hacking a radio-controlled car, for example. You can easily rework the truck example from Chapter 8 to use the WiFly instead of a RedBearLab BLE Shield.

With the basics of WiFi communication mastered, it's time to put this to use with a few fun hacks. The next chapter shows just a few of the exciting things you can do with a WiFly.

WiFi Servos

Servos: They're Where the Action Is

One of the most versatile mechanical devices available for robotics, or pretty much any kind of physical control work, is the *servo* (shown in Figure 12-1). A servo is a package containing a motor and electronics that responds to a signal by rotating to a specified location. Give it one value, and it will spin to the left; another spins to the right. Simple mechanical linkages (aka horns) can turn this motion into a push-pull motion, which is how servos control radio-controlled model airplanes. Servos are generally limited to about 180 degrees of motion, but they can easily be hacked to rotate 360 degrees. Since the servo can still be turned a specific amount, the hacked version is frequently used for controlling wheels on small robots.

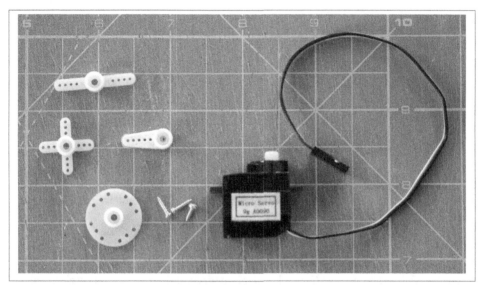

Figure 12-1. A servo and control horns

The actual control signals for servos can be a bit complicated, but fortunately, servo controllers are readily available that do the heavy lifting for us. Pass them a few numbers and the servo will turn to a specific position, and even do it at a specified speed. Of course, you need a way to pass the numbers. As it turns out, you learned one good way in the last chapter.

We're going to build two servo projects in this chapter. The first uses the back and forth turning motion of a servo to rotate an eyeball. It's going to be perfect at Halloween, tucked into the head of a Jack-o'-lantern! The second will show how to use a servo to get a push-pull motion. In this case, we'll remotely control an M&M dispenser, causing it to dump candy on command.

The specifics of the projects are not as important as learning to use servos in general. The chapter ends with some suggestions for other things you could do with the servos.

Table 12-1 provides a list of the necessary parts for the project.

Table 12-1. Parts list

Part	Description
Any iPhone, iPad, or iPod	They all have WiFi.
techBASIC or techBASIC Sampler	We'll use the WiFly Terminal program, preinstalled in techBASIC.
WiFly	More formally, the RN-XV WiFly Module with Wire Antenna, or WRL-10822. This is the same part you used in the last chapter.
XBee Breakout	Again, this is the same part used in the last chapter. If you used a substitute there, it will work fine here, too.
2 mm 10-pin XBee Header (2)	You will need two 1×10 headers with a 2 mm spacing for the XBee Breakout board. These are an unusual spacing, so plan ahead—you can't pick these up at most local electronics stores.
Break Away Headers - Straight	You will also need two 1×10 headers with 0.1″ spacing for the XBee Breakout board. These are the common headers you can find almost anywhere. This specific part is from SparkFun (*http://www.spark fun.com*), and is a 1×40 header you can clip apart to any desired length.
DFRobot Breadboard Power Supply 5V/3.3V	This power supply is particularly nice for the servo projects, since it provides both 3.3V and 5V (the WiFly needs 3.3V and the servos need 5V). If you're using a substitute, be sure you have both voltages available.
Breadboard and jumpers	You'll need a breadboard and some jumper wires for connecting everything. An alternative is to mount the components on a custom circuit board. That will be necessary for some projects, but we can get by with breadboards for this chapter.
Pololu Serial Servo Controller	Takes commands from a serial line and converts them to control signals for up to eight servos.
Servo	Any small servo will do. I used the ROB-09065 RoHS from SparkFun.
Piano wire	Thin and stiff, piano wire is a great choice for making servo linkages. I used 0.039″ diameter wire. Visit the hobby store once you pick your project, though. There are lots of alternatives for various purposes.
Control horn	This is a piece used at the end of a control rod that attaches to the object that needs to move. In this chapter, it's the lever on an M&M dispenser.
Hackable objects	You'll need something to manipulate. Be creative! Anything that can operate with 180° turning motions or a push-pull motion of a half-inch or so can be hacked for servo control.

The Pololu Serial Servo Controller

The Pololu Serial Servo Controller takes the serial signals from the WiFly and controls up to eight servos. We're going to start with one servo, but you'll see that it is really easy to add more, so you might want to order a couple of spare servos. As with all small electronic components, it's not a bad idea to order a couple of servo controllers, too. You're either very skilled or very lucky if you haven't burned out a component yet. I burned out one of the WiFly boards myself.

You can buy the servo controller preassembled or as a kit. The kit already has all of the surface mount components installed, as shown in Figure 12-2. All you have to solder are the headers. As you'll see in a moment, there are advantages to the kit if you are using a breadboard.

Figure 12-2. The parts in the Pololu servo controller kit

The Pololu website has instructions both for construction and for the signals the controller expects. We'll cover everything you need to know for this project here, but it's still a good idea to get a copy of the official User's Guide (*http://www.pololu.com/file/ 0J37/ssc03a_guide.pdf*).

Solder the 2×9 header first. This is the one that supplies power for the servos. Tack it down by soldering a pin on one of the outside corners, then check to make sure everything is aligned. Solder the opposite inside corner and check again. If everything is still aligned properly, solder the other 16 connections.

You'll need four other headers: a 1×8, a 1×5, and two 1×2 headers. The kit comes with a 1×13 and a 1×4, so you'll need to clip the headers apart with a pair of diagonal cutters, as shown in Figure 12-3.

Solder the 1×8 header next. This time, don't just check the alignment visually: try inserting the three-pin header connector on the servo on the connectors for servos 0 and 7, just to make sure they are aligned. Tack each end, checking the alignment both times, then solder the remaining six posts. Follow up with the two 1×2 headers, one for the comm connection and one for the mode jumper.

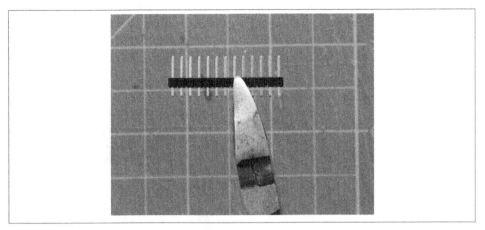

Figure 12-3. Separating the headers

You have a choice when you get to the 1×5 jumper. This is the one that will be used for most of the connections to the circuit we will build. The instructions for the part show this header pointed up, like the others. For our circuit, it's easier to point it down as shown in Figure 12-4, so it can plug directly into the breadboard. You can do it either way, but if you point the header up, be prepared to build a 1×5 connector, tack-solder some wires to the header, or use jumper wires to make the connections to the breadboard.

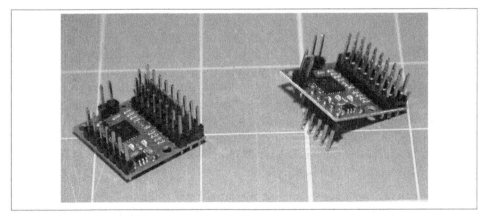

Figure 12-4. Finished servo controller—install the five-pin header with pins down for breadboard use

The Circuit

If you installed the 1×5 header upside down, plug the board into the breadboard as shown in Figure 12-5. Add a piece of tape to the bottom of the circuit if it appears the header pins are sticking down far enough to make contact with the connectors in the breadboard. You might want to fasten the servo controller to the breadboard with a small piece of mounting tape if you installed the 1×5 header pointing up, just to keep things stable.

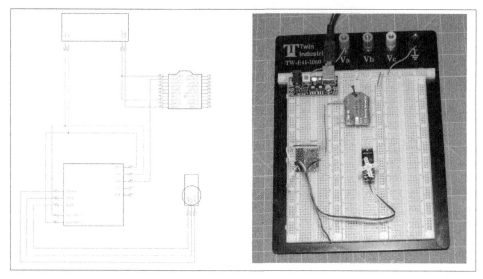

Figure 12-5. Servo circuit

The servo controller has two separate connections for 5V. The one on the 1×5 header powers the board itself, while the one at the top of the 2×9 connector is used to power the servos. For a single small servo with light loads like the one we're using, both sides of the circuit can be powered from the DFRobot power supply, but things change with larger servos, lots of servos, or heavy motor loads. You may need up to 10 amps with some servo combinations, and that's enough power that the servos would need their own power supply—so the servo controller gives you that option. Connect both to the 5V side of the DFRobot power supply. If you put the 1×5 header in upside down, that side of the circuit is easy. You'll need to use jumper wires or a 1×2 connector to connect the servo power to the breadboard.

There are actually two ways to connect the serial lines of the two devices. The servo controller conveniently provides both a standard RS-232 serial port and one labeled *logic-level serial input*. The logic-level connection is perfect for connecting to the WiFly. Connect that pin on the servo controller to pin 2 of the WiFly, which is TX, the serial transmission pin.

The servo itself has a 1×3 connector that can plug into any of the servo connections. The stock software is set up to control servo 0, so it's easiest to plug the servo into that connection. Be sure the white wire is on the inside of the board and the black wire on the outside, as shown. Mount the servo to the breadboard with a small piece of mounting tape.

So we're done, right? Well, almost. The servo controller is a bit picky about the signals it gets, and will go into an error mode if it gets unexpected, random input. Unfortunately, the first thing the WiFly does is spit out some unexpected output to signal that it is awake. That is going to put the servo controller in an error mode, with a solid red LED and a flashing green one. There are several ways to handle this, but by far the simplest is to bring the reset pin high by connecting it to +5V for an instant. You can get fancy and do that with a switch. Or you could get even fancier, and wire up the WiFly to send a signal using a PIO pin and, probably, an intervening NMOS logic chip to bring the 3.3V signal up to 5V. Or you could just dangle an extra wire off of the reset pin and touch it to +5V when the serial controller needs to be reset. That's what I ended up doing.

Halloween Hijinks

A creative and slightly warped mind will come up with all sorts of fun things to do with a device that can turn something back and forth by 180 degrees. I decided to turn an eyeball to get ready for Halloween pranks. It's remarkably easy.

A ping pong ball and a few minutes with some colored markers will create a convincing eyeball. Use a small dab of superglue or mounting tape to tack it to the round connector that comes with the servo kit. Don't worry about the screw—the eyeball is very light, so simply pressing the connector onto the servo will hold it well enough.

You might have other ideas about what to do with the servo. Be creative. Remember, anything that turns is a candidate for the servo, from a bobble-head to a robot part.

The Software

The Pololu Serial Servo Controller actually supports two different protocols for controlling the servo. You select the command set using the blue jumper pin, sliding it onto the Mode pins for Mini SSC II Mode, and leaving the blue jumper pin off for the Pololu command set. We'll use the Pololu command set, since it gives the option of controlling the speed. Who wants an eyeball to rotate quickly? Slow is much more spooky.

Figure 12-6 shows what the software looks like when it is running. We'll use this as a guide while walking through the code. As always, the source is in the *O'Reilly Books* folder in techBASIC and techBASIC Sampler. The program is called Serial Servo.

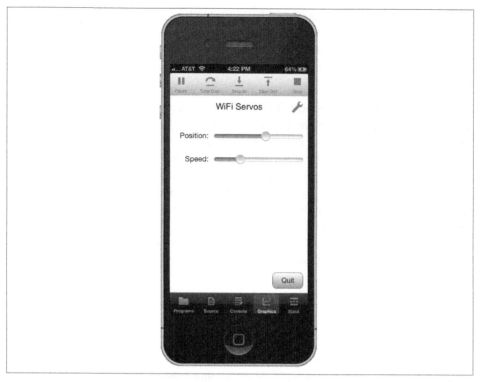

Figure 12-6. GUI for the servo software

```
! This app uses an RN-XV WiFly and a Pololu Micro Serial Controller
! to control the position and update speed of a servo. The servo
! controller can control up to 8 servos, and the app can easily
! be extended to do the same.

! Declare the controls.
DIM positionSlider AS Slider, speedSlider AS Slider, quitButton AS Button
```

If you've followed along through the rest of the book, this is pretty standard stuff, so we'll go over it quickly. The DIM statement declares variables for the three interactive controls in the GUI so they can be accessed from multiple subroutines.

```
! Define the device and servo numbers.
DIM deviceID AS BYTE, servoNumber AS BYTE
deviceID = 1
servoNumber = 0
```

The servo controller has a pretty cool characteristic: you can daisy-chain several together for projects like a walking robot, which might use quite a few more than just eight servos. The default device number is 1, but the program puts the number up front so it is easy to find and change. The same is true for the servo number. This corresponds to the pin position on the servo controller, and should be a number from 0 to 7.

It is critically important that the variables are defined as BYTE, not INTEGER or, heaven forbid, the default type of SINGLE. We'll see why in a moment.

```
! Open a channel to the servo controller.
Comm.openTCPIP(1, "169.254.1.1", 2000)
```

Take a look back at the previous chapter if you need some clarification here. This just opens the TCP/IP connection to the WiFly.

```
! Set up the user interface.
setUpGUI
```

This calls a subroutine to set up the user interface.

```
! Set up the user interface.

SUB setUpGUI
DIM positionLabel AS Label, speedLabel AS Label, nameLabel AS Label
nameLabel = Graphics.newLabel(10, 10, 300, 30)
nameLabel.setText("WiFi Servos")
nameLabel.setAlignment(2)
nameLabel.setFont("Arial", 20, 0)

y = 80
positionLabel = Graphics.newLabel(10, y, 80)
positionLabel.setText("Position:")
positionLabel.setAlignment(3)

positionSlider = Graphics.newSlider(100, y, 210)

y = y + 53

speedLabel = Graphics.newLabel(10, y, 80)
speedLabel.setText("Speed:")
speedLabel.setAlignment(3)

speedSlider = Graphics.newSlider(100, y, 210)
speedSlider.setValue(1)

quitButton = Graphics.newButton(Graphics.width - 82, Graphics.height - 47)
quitButton.setTitle("Quit")
quitButton.setBackgroundColor(1, 1, 1)
quitButton.setGradientColor(0.6, 0.6, 0.6)

System.showGraphics
END SUB
```

Setting up the user interface is pretty standard stuff by now. We set up the three labels, one with the program name and two to label the sliders, a Quit button, and the two sliders.

```
! Handle a tap on a button.
!
```

```
! Parameters:
!    ctrl - The button that was tapped.
!    time - The time stamp when the button was tapped.

SUB touchUpInside (ctrl AS Button, time AS DOUBLE)
IF ctrl = quitButton THEN
  STOP
END IF
END SUB
```

touchUpInside is set up to stop the program if the user presses the Quit button.

```
! Handle a value changed event.
!
! Parameters:
!    ctrl - The control whose value changed.
!    time - The time stamp when the change occurred.

SUB valueChanged (ctrl AS Control, time AS DOUBLE)
```

Here's the interesting stuff. When one of the sliders is moved, we need to send a command to the servo controller. The program starts by checking to see if the slider that changed was the position slider—the one that controls the direction the servo turns.

```
IF ctrl = positionSlider THEN
  ! The position slider changed. Move the servo.
  b~ = 128 ❶
  PUT #1,,b~
  PUT #1,,deviceID ❷
  b~ = 4 ❸
  PUT #1,,b~
  PUT #1,,servoNumber ❹
  position% = ctrl.value*5000 + 500 ❺
  b~ = position% >> 7 ❻
  PUT #1,,b~
  b~ = position% BITAND $007F ❼
  PUT #1,,b~
```

If the slider was the position slider, here's what the code does:

❶   There are several commands the program can send to the servo controller. All commands start with a single byte with the number 128, or hexadecimal $80. This value must be stored in a byte variable, not an integer or some other type. We'll use the PUT command to send the value to the WiFly, and the PUT command will send two bytes if the variable is an integer. Unless declared otherwise in a DIM statement, variables whose names end with ~ are byte variables in BASIC.

❷   The next byte is the device ID. This allows the use of multiple devices. If you recall, deviceID was set to 1 at the start of the program.

---

❸ The next byte sent to the WiFly must be the command number. There are several ways to move a servo, depending on whether you want to move it to an absolute location or just a certain distance from the position it is already at. The slider motif works best with the absolute position, so the program passes the number 4, indicating it will pass an absolute position.

❹ Next is the servo number. Again, this was set at the start of the program.

❺ The servo controller expects the position of the servo as a number from 500 to 5,500. This line grabs the slider position and converts it to a value from 500 to 5,500, placing the value in a two-byte integer. The % character tells the compiler the value is an integer, just as ~ indicated a byte.

❻ A single byte can only represent numbers from 0 to 255, so we'll need more bits. Unfortunately, the servo controller uses the most significant bit of a byte to indicate the start of a command, so it's not as simple as just writing a two-byte integer value, since the most significant bit might be set. The >> 7 operation shifts the bits right by seven bit positions, dumping these bits and leaving the most significant bits of the original value. ctrl.value ranges from 0 to 1, so the largest value position% will be set to is 5500. We need 13 bits to represent 5500, so the resulting value will use at most 6 bits, leaving the most significant bit set to 0.

❼ Finally, the program uses the BITAND $007F operation to mask off the high bits, leaving just the least significant seven bits. These are sent in the last byte of the command.

```
ELSE IF ctrl = speedSlider THEN
  ! The speed slider changed. Change the movement
  ! speed for the servo.
  b~ = 128
  PUT #1,,b~
  PUT #1,,deviceID
  b~ = 1
  PUT #1,,b~
  PUT #1,,servoNumber
  b~ = ctrl.value*127
  IF b~ = 0 THEN b~ = 1
  PUT #1,,b~
END IF
END SUB
```

The speed slider controls how fast the servo swings from one position to another. It takes values from 1 to 127, setting the speed of the servo to somewhere between 50 microseconds of arc per second of time at the low end and 6.35 milliseconds of arc per second of time at the high end.

Other than changing the command number from 4 to 1 and writing a single byte instead of two to set the position, this section of the code is pretty similar to the section that handled the servo position.

Take It for a Spin

Well, what are you waiting for? Use the Settings app to connect your iOS device to the WiFly, just like you did in the previous chapter. Then crank up Serial Servo from the *O'Reilly Books* folder. If everything is wired properly, the position slider should spin the eyeball around convincingly (Figure 12-7).

Figure 12-7. Spooky servo rotation

It will start at full speed, but once you move the speed slider, the speed will slow down a bit. The servo may not be capable of really rapid movement, though, so you may only see a noticeable change in speed when the speed slider is at the low end of its range.

Push and Pull with Servos

The various plastic parts that come with the servo are designed to convert a rotational motion into a push or pull motion. They are called *control horns*. There is almost always a corresponding control horn attached to the object to be pushed or pulled. These are connected with a *control rod* or *control cable*.

Control rods are made with stiff wire called piano wire or music wire. It's available from many hobby stores, especially the ones that carry radio-controlled airplanes. Control rods are the typical method for controlling gas- and electric-powered radio-controlled airplanes. A control cable generally takes the form of a plastic tube that provides stiffness for a thin twisted steel cable. Again, it's available from hobby stores that carry model airplane parts. Control cables are the most common way to connect control surfaces in radio-controlled sailplanes.

The control horn at the top left of Figure 12-8 is part of the servo kit. This time it's probably best to attach the horn to the servo with a screw rather than depending on a pressure fit, but you can try the pressure fit first. The control horn at the top right is from a hobby store. It's a very common style for small airplanes. Unfortunately, the base is rotated 90 degrees from what I needed, so I used a small piece of scrap balsa to provide a base pointed in the direction needed.

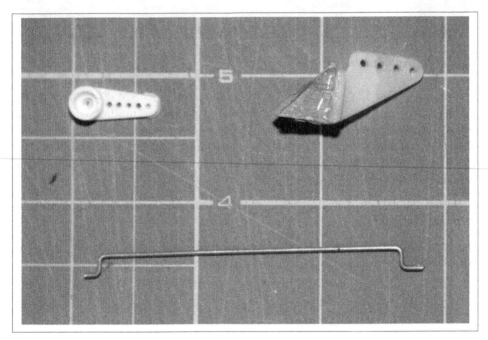

Figure 12-8. Controls and linkages

Control rods can be very simple or very complicated, depending on the size of the servo, the length of the rod, and how much force the rod will endure. The one shown in Figure 12-9 is made from 0.039″ piano wire, bent to fit the specific toy I hacked. Bend the wire using needle-nose pliers, then cut it to length using a pair of diagonal cutters.

Figure 12-9. Installed linkages and servo

For my application, I'm pushing a lever that will control an M&M dispenser, spewing M&Ms from the mouth of a nutcracker. The servo is mounted on the back of the nut-cracker using mounting tape—that thick, double-sided tape that looks red because a red plastic strip separates the sticky parts. It's pretty strong, and just flexible enough to make the entire assembly a little forgiving of tolerances. I used super glue to fasten the wood attached to the control horn to the lever.

The electronics and software are all identical to that used for the rotating eyeball, so we're good to go.

Figure 12-10 shows the nutcracker in action. The side view clearly shows how the control horns work together to press the lever down, opening the mouth.

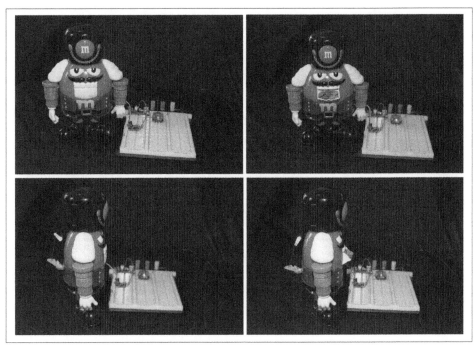

Figure 12-10. The nutcracker in action

While a nutcracker was used here, you can use this circuit to control pretty much any small device that operates with a push-pull motion and doesn't require too much force. That could be a doorbell, an automatic pet feeder, or lots of toys.

Pomp and Circumstance

Combining the tools and techniques from this chapter with the ones we looked at earlier, you have at your fingertips an impressive collection of hardware, software, and electronics techniques that can be applied to any number of situations, from controlling toys to collecting field data for the physical sciences. All of these techniques can be adapted to other projects, from robotics to weather stations, pranks to product prototyping. I'd love to hear of the fun and creative things you do. Drop by the book's O'Reilly page (*http://shop.oreilly.com/product/0636920029281.do*) or the Byte Works Facebook page (*https://www.facebook.com/ByteWorks*) and let us and your fellow readers see what you've done to control your world!

Index

We'd like to hear your suggestions for improving our indexes. Send email to index@oreilly.com.

About the Author

Mike Westerfield started programming on a PDP-8 using a teletype terminal. As the personal computer revolution got going, he sold his car and rode a bike for several months to raise cash to buy an Apple II computer. He wanted to write a chess program but couldn't find a good assembler, so he took a summer off to write his own. Two years later he finished ORCA/M, which went on to become Apple Programmer's Workshop, the Apple development environment for the Apple IIGS.

Born the same year as Steve Jobs and Bill Gates, Mike made the mistake of getting an education instead of getting rich. A slow learner, he graduated from the US Air Force Academy in 1977 with a degree in physics, earned an MS in physics from the University of Denver, and was working on a PhD when he started making more money from his sideline software company than he did from the Air Force.

Since then Mike has developed numerous compilers and interpreters, software for mission-critical physics packages for military satellites, plasma physics simulations for Z-pinch experiments, multimedia authoring tools for grade schoolers, disease surveillance programs credited with saving the lives of hurricane Katrina refugees, advanced military simulations that protect our nation's most critical assets, and technical computing software for iOS.

Mike currently runs the Byte Works, an independent software publishing and consulting firm. He is a PADI scuba instructor who lives in Albuquerque with his wife, where he enjoys being an empty nester and spoiling his grandchildren.

Colophon

The animal on the cover of *Building iPhone and iPad Electronic Projects* is a Eurasian magpie (*Pica pica*). The magpie is a very common bird throughout Europe, much of Asia, and northwestern Africa. Although considered a pest by some due to its penchant for eating small songbirds, the magpie is one of the smartest species of birds, and indeed one of the smartest animals in general. In the wild, magpies have been observed engaging in elaborate social rituals; they use tools, hide and store food, and employ complex group hunting strategies. Other members of the magpie's family, *corvidae*, include crows and ravens, and all of these species have exhibited mirror self-recognition in captivity. Some primates and cetaceans share this capability; these animals have a large brain-to-body weight ratio that is only slightly lower than that of humans. It has even been shown in laboratory tests that the brains of *corvidae* birds have evolved the same ability to think geometrically as the great ape's.

Wild magpies form monogamous pairs after attending large gatherings that Charles Darwin described as "marriage meetings." A mating pair of magpies will stay together for their entire lives, raising broods of five to eight chicks every year. The chicks stay in the nest for a few weeks after hatching and still remain with the parents for about a week after learning to fly. It is thought that this long period of adolescence helps contribute to the birds' intelligence, since the chicks have ample time to learn social behaviors from their parents.

The name "magpie" is derived from a mention in Shakespeare's *Hamlet* of a "magot pie," or "pied Margot." A pied (or piebald) animal is one that has a spotting pattern of large white areas over black; in the play, Hamlet describes this bird along with "choughs and rooks" who bring forth "the secret'st man of blood." The magpie has long been entwined with European folklore, and even in Shakespeare's time the bird was looked upon as a dark omen. Most of the myths associated with the magpie are a result of the bird's tendency to "steal" shiny objects or of its aggressive behavior toward songbirds. In some areas of Britain, it is still traditional to greet a solitary magpie with "Hello, Mr. Magpie, how is your wife today?" to ward off misfortune.

The cover image is from Wood's *Natural History*. The cover font is Adobe ITC Garamond. The text font is Adobe Minion Pro; the heading font is Adobe Myriad Condensed; and the code font is Dalton Maag's Ubuntu Mono.

Have it your way.

CPSIA information can be obtained at www.ICGtesting.com
Printed in the USA
BVOW10s1308200913

331172BV00004B/2/P